Forest and Plant Bioproducts Division

Presentations at the 2009 AIChE Annual Meeting

Nashville, Tennessee, USA
8-13 November 2009

ISBN: 978-1-61567-939-3

Printed from e-media with permission by:

Curran Associates, Inc.
57 Morehouse Lane
Red Hook, NY 12571

Some format issues inherent in the e-media version may also appear in this print version.

Printed by Curran Associates, Inc. (2010)

For permission requests, please contact AIChE
at the address below.

AIChE
3 Park Avenue
New York, NY 10016-5991

Phone: (203) 702-7660
Fax: (203) 775-5177

www.aiche.org

Additional copies of this publication are available from:

Curran Associates, Inc.
57 Morehouse Lane
Red Hook, NY 12571 USA
Phone: 845-758-0400
Fax: 845-758-2634
Email: curran@proceedings.com
Web: www.proceedings.com

TABLE OF CONTENTS

Surface Functionalisation of Bacterial Cellulose Nanofibrils as the Route to Produce Green Polylactide Nanocomposites with Improved Properties

Koon-Yang Lee, Department of Chemical Engineering, Imperial College London, London, United Kingdom

Jonny J. Blaker, Chemical Engineering, Imperial College London, London, United Kingdom

Alexander Bismarck, Chemical Engineering, Imperial College London, London, United Kingdom

Recent interest in environmental friendly products and 'the public's growing demand for greener materials has sparked the use of cellulosic materials as the reinforcement in composites. This is not surprising as cellulosic materials such as natural fibres have distinct advantages over conventional glass fibres; low cost, low density, high toughness and most importantly, biodegradability. In addition to natural fibres, bacterial cellulose is also one of the extensively studied reinforcement. Bacterial cellulose possesses a highly crystalline structure (~90%) and a high Young's modulus of about 114 GPa. This value is comparable to that of high performance glass fibres considering that bacterial cellulose has much lower density (1.25 g cm-3) than glass fibres (2.5 g cm-3). It also exists as a nano-sized material naturally and possesses a high surface area of about 150 m2 g-1. All these properties are highly favourable for the use of composite production. However, bacterial cellulose is extremely hydrophilic in nature and it has strong affinity to itself due to the presence of large amount of hydroxyl groups (-OH) on the surface. This will often result in poor interfacial adhesion between bacterial cellulose and hydrophobic polymer matrices, which limits the application of bacterial cellulose in the composites field.

Therefore in this work, we investigate the effect of surface functionalised bacterial cellulose with long chain fatty acid (acetic acid, hexanoic acid and lauric acid) for the use in polylactide based nanocomposites. By limiting the reaction only to the surface of bacterial cellulose, the highly crystalline structure will be retained whilst the cellulose surface will become hydrophobic. This will lead to an improvement in the interfacial adhesion between bacterial cellulose and polylactide. We have explored a method based on thermally induced phase separation to ensure the processability of dried bacterial cellulose in an extrusion process. A method was also developed to determine the contact angle of polylactide droplets on (modified) bacterial cellulose nanofibrils. We could show a significant improvement of the wettability of bacterial cellulose nanofibrils by polylactide with increasing hydrophobicity of the nanocelluloe fibrils, i.e. the longer the chain length of the grafted fatty acid. This indicates a much improved adhesion between the modified cellulose and polylactide. The tensile modulus and tensile strength of the resulting polylactide nanocomposites were found to increase by as much as 50% and 15%, respectively with bacterial cellulose loading fractions of only 5% compared to neat polymer. The thermal degradation and the viscoleastic behaviour of the polylactide nanocomposites were also found to improve as well. This implied that surface functionalisation of bacterial cellulose nanofibrils can be used as the route to produce truly green nanocomposites with improved properties.

See more of this Session: Biobased Materials - I
See more of this Group/Topical: Forest and Plant Bioproducts Division

All Green Composites for a Sustainable Manufacturing: Where We Are and Future Directions!

Amar K. Mohanty, Department of Plant Agriculture & School of Engineering, Guelph, ON, Canada

Biobased economy is challenging to agriculture, forestry, academia, government and industry. The incorporation of bio-resources, e.g. crop-derived green plastics and plant derived biofibers (natural fibers) into composite materials are gaining prime importance in designing and engineering green composites. Biocomposites derived from natural fibers and traditional polymers like polypropylene, polyethylene, epoxy and polyesters have been developed for automotive parts and building structures. Renewable resource based bioplastics like polylactic acid (PLA), polyhydroxyalkanoates (PHAs), biobased polytrimethylene terephthalate (PTT), cellulosic plastics, soy/corn/wheat protein based bioplastics and vegetable oils derived bioresins need value-added and diverse applications to compete with the fossil fuel derived plastics. Through reactive blends, composites and nanocomposites new biobased materials are under constant development. The door is opening and path is clearing-up for many emerging biopolymers and biobased composite materials that are poised to create a major break through in the commercial in-roads.

Natural fibers are lighter, less expensive, have superior specific strength, require comparatively less energy to produce, are good for the environment, biodegradable and have superior sound abatement characteristics as compared to synthetic glass fibers. All of these attributes are quite favorable, especially in the automotive sector where even a fractional weight saving can make a significant contribution to energy savings with reduced gasoline consumption and with added advantages of eco-friendliness. It is true that natural fibers are comparatively hydrophilic and less thermally stable as compared to glass fibers. However, the recent developments of natural fiber technologies overcome these disadvantages if used intelligently.

Hybrid and intelligently engineered green composites are going to be the major drivers for sustainable developments. Besides agricultural natural fibers like kenaf, jute, flax, industrial hemp, sisal and henequen; inexpensive biomasses such as wheat straw, rice stalks, corn stovers, grasses, soy stalks and lignin (the byproducts from pulp and paper and lingo-cellulosic ethanol industries) have great potential for use in sustainable biobased composite materials.

This presentation will highlight the current status, opportunities and challenges of bioplastics, natural fiber composites and all green composites for uses in car parts, consumer goods and sustainable packaging.

See more of this Session: Biobased Materials - I
See more of this Group/Topical: Forest and Plant Bioproducts Division

Sustainable Bio-based Green-Composites for Automotive Interior Parts

Hyun-Joong Kim[a,b], *Byoung-Ho Lee*[a]

[a] *Laboratory of Adhesion and Bio-Composites, Program in Environmental Materials Science,*
[b] *Research Team for Biomass-based Bio-Materials, Research Institute for Agriculture and Life Sciences, Seoul National University, Seoul, Republic of Korea.*
www.adhesion.org (hjokim@snu.ac.kr)

Introduction

Advanced polymer composites containing carbon and glass fibers have been utilized extensively in the aerospace, automotive, and construction industries. Since the matrices and the fiber reinforcements in these advanced composites are based on mineral resources, their long term sustainability is problematic. While recycling may be a viable strategy, the complicated mixed morphology of composite materials makes them inherently difficult to recycle. In comparison, several so called green-composites, have been developed that offer certain environmental advantages at the end of their use cycle when composites are landfilled or incinerated. For the purposes of this study, green-composites are defined as composite materials that combine natural fibers such as sisal, jute, hemp, and kenaf with either biodegradable or non-biodegradable polymers. Natural fibers have many advantages over synthetic fibers; these advantages include biodegradability, low density, high toughness, acceptable specific strength, reduced dermal and respiratory irritation, low cost, and less use on non-renewable resources.

Recently, 'Sick Car Syndrome' was occurred by the problem for new made cars that have great quantity VOCs (volatile organic compounds) in their interior parts. Thus, the automotive makers have struggled to reduce emitted VOCs from car interior. In building fields, it had researched TVOC reduction method, such as ventilation equipment, bake-out treatment. A bake-out has been used to reduce VOC emissions from newly installed materials, products, and furnishings. This word in usually used to prevent building-associated illness caused by the off gassing of volatile organic compounds (VOC) and formaldehyde from residual solvents in new building materials and furnishings.

Therefore, in present study, we investigate bio-composite with sustainable reinforced natural filler to enhance thermal and mechanical properties of green-composites. And manufactured bio-composites were experimented by thermal test such as HDT, thermal aging test for application of car interior parts. And TVOC emission behavior from bio-based green-composites materials was investigated for the solution of TVOC reduction in car indoor by various test methods.

Experimental

Materials

PLA was supplied by Cargill-Dow Co. USA with an MFI of 15g/10min (190°C/2,160g) and a density of 1.22 g/cm^3. The maleic anhydride-PLA (MAPLA) was synthesized at Seoul National University by reactive extrusion at 185□ and 200 rpm, using PLA, maleic anhydride (Merck, Whitehouse Station, NJ, USA) and dicumyl peroxide (Aldrich, St. Lousis, MO, USA) as a coupling agent in the proportion 100:1:0.1 w/w. PBS was prepared at Ire Chemical Ltd., Korea, with an MFI of 25g/10min and a density of 1.26 g/cm^3. The natural flours used as the reinforcing filler, pineapple skin flour and destarched cassava root flour were supplied by CSTRU (Cassava and Starch Technology Research Unit) in Thailand. The second bio-flour was prepared at the CSTRU by treating non-destarched cassava root flour with α-amylase (2 hour, 100□) and glucoamylase (overnight, temperature less than 60□) to remove the starch fraction (up to 69% dry weight basis; Sriroth *et al.*, 2000), resulting in destarched cassava root flours.

Compounding and sample preparation

Pineapple flour and destarched cassava flour were oven dried at 105□ for 24 hrs to adjust the moisture content to 1~3% and then stored in sealed polyethylene bags before compounding. The PLA was blended with each flours and MAPLA in a laboratory-sized, co-rotating, twin screw extruder (model BA-19, Bau Technology, South Korea) using three general processes: melt blending, extrusion and pelletizing. The extruder barrel was divided into eight zones with the temperature in each zone being individually adjustable. Also, to enhance interfacial adhesion, the 30 wt% filler loading was prepared by incorporating 3% MAPLA. The temperature of the mixing zone in the barrel was maintained at 185□ with a screw speed of 250 rpm. The extruded strand was cooled in a water bath and pelletized using a pelletizer. The extruded pellets were oven dried at 80□ for 24 hrs and stored in sealed polyethylene bags to avoid unexpected moisture infiltration.

Heat deflection temperature (HDT)

The heat deflection temperature is measurements of bio-composite's resistance to distortion under a given load at elevated temperatures. The deflection temperature is also known as the 'deflection temperature under load' (DTUL) or 'heat distortion temperature' (HDT). It is used to determine short-term heat resistance. It distinguishes between materials that are able to sustain light loads at high temperatures and those that lose their rigidity over a narrow temperature range. HDT analysis of bio-composites is measured by a Toyoseiki HDT tester, and followed the ASTMD 648-98 method. The loading pressure is 0.455 Mpa (66 psi) and heat at the rate of 2□/min from room temperature to 150□. The test specimen size of HDT is 127 × 13 × 3 (mm^3).

Thermal aging test

Thermal aging test was used to confirm for the vehicle interior application of manufactured bio-composites by chamber (Weiss Umwelttechnik GmbH Simulationsanlagen, WEISS TECHNIK). The tensile samples were prepared for thermal aging test that was set -20, 0, 50 and 80°C each temperature was maintained 15 minute. Figure 1 shows the temperature schedule of manufactured samples that had 200 cycles and five samples were selected each 50 cycle for shrinkage ratio, tensile and flexural test. Five measurements were conducted and averaged for the final result. To shrinkage ratio was calculated according to the following equation:

$$R = \frac{100(L_0 - L_a)}{L_0}$$

where R is the shrinkage ratio of the PLA bio-composites and matrix polymers, L_o is the length of the first specimens, and L_a is the length of the thermal aging treated specimens at each cycles.

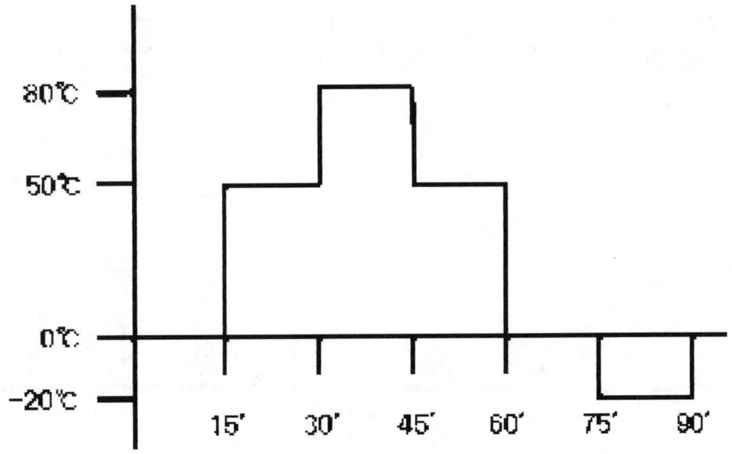

Figure 1. The temperature schedule for thermal aging test of bio-composite.

Thermal extractor analysis (TE)

The TE analysis is mainly used to measure emitted TVOC (Total Volatile Organic Compound) and formaldehyde from construction materials, MDF (Medium Density Fiberboard), PB (Particle Board), paints and adhesives. At first, to prepared sample for TE, extruded pellets were injection molded into test bars using an injection molding machine (Bau Technology, South Korea). The test specimen size of TE is 47 × 6 × 1.5 (mm). After injection molding, the test bars were conditioned before testing at 50±5% RH for at least 24 hours and sealed using aluminum foil. The temperature of bio-composites was heated in glass tube surrounded in heating oven from 30°C to 90°C by Gerster Aux Controller 163. The air-flow of TE was controlled at 39 mL/min using high purified nitrogen gas by Gerstel Airflow pneumatic and 1L of TVOC was sampled. This gas was introduced into the inlet of the glass tube and contained emitted VOCs from bio-composites. VOCs (Volatile Organic Compounds) of bio-composites (PLA bio-composites and PBS bio-composites) in the glass tube were sampled by Tenax TA tubes. The VOC concentrations were analyzed by a gas chromatography-mass spectrum detector (GC-MSD).

Results and discussion

Heat deflection temperature (HDT)

HDT measures thermal properties of composites, such as the thermal sensitivity during relatively short-term. To add bio-flour in composite is the most positive effects aspect of the increase HDT. The HDT of the WF-reinforced PLA composites is higher than the HDT of the PLA resin, and this value suggests the upper service temperature limit for a plastic [Huda *et al.* 2006].

Figure 2 shows the HDT of the PLA-composites with the 30% flour content with and without compatibilizer. As the loading pineapple flour and destarched cassava flour the HDT of the PLA-composite increases from 56.8°C to about 66.3°C and to about 69.7°C, respectively. This may be better transfer stress in PLA bio-composite with destarched cassava flour than with pineapple flour, because destarced cassava flour had a size distribution relatively shorter length in PLA-composite than the length of pineapple flour. And the HDT of PLA bio-composite with pineapple flour with MAPLA and PLA bio-composite with destarched cassava flour with MAPLA is 70.6°C and 82.2°C, respectively. Because better chemical bonding exists between the PLA matrix and flours (pineapple flour and destarched cassava flour), the PLA bio-composites with compatibilizer reveal a higher HDT than composites without compatibilizer. PLA-C composite with compatibilizer (an increase of about 17.9%)

compare with PLA-P composite with compatibilizer (an increase of about 6.5%) reveal the higher HDT ratio, because increasing the crystallinity and reinforcement are generally the important factors to increase the HDT of composite.

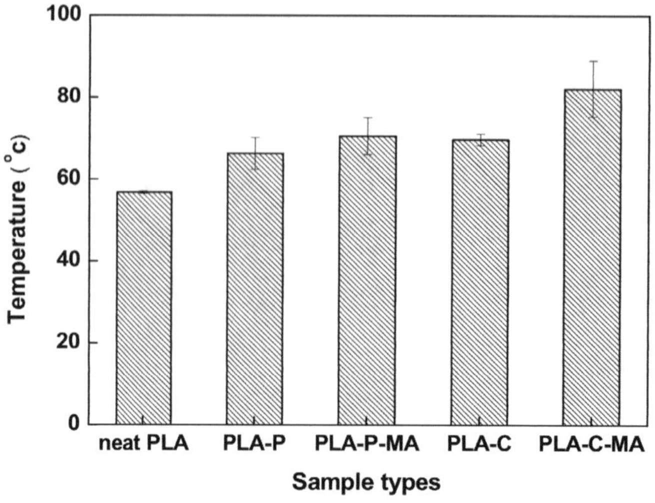

Figure 2. HDT of bio-composites.

Thermal aging test

All most of polymers employed mainly for the protection of outdoor environment, interior materials, necessities of life and so on. These polymers are needed suitable properties for their use, aim and life time. So, polymers should have guaranteed by various weathering test, such as salt spray, photochemical and marine atmosphere, UV, and natural exposure [Favaro *et al.* 2008]. We investigated to apply for car interior using thermal aging test, which is scheduled according to car cabin temperature variation, which was set -20°C, 0°C, 50°C and 80°C considering season and regions of Korea.

As shown in Figure 3, the tensile strength of neat PLA maintained consistently from 50 cycles to 200 cycles as 62.3 MPa and 60.1 MPa, respectively. Although the weak bonding between PLA molecular by heat impact, it seems that the maintain strength for increasing crystallinity of neat PLA during thermal aging teat that crystallinity of PLA increased by degradation [Duek *et al.* 1999, Weir *et al.* 2004]. Whereas, tensile strength of PLA bio-composites with compatibilizer or without compatibilizer rapidly was decreased, that had lower value than tensile strength of polypropylene which revealed stable strength around 40 MPa. The crystallinity of annealed PLA matrix decreasing and the wider gap between PLA matrix and flours for shorten molecular of PLA is considered and MAPLA is not effect as compatibilizing agent in thermal aging test. Because MAPLA is composed of PLA matrix, it is considered same properties compared with neat PLA. However, in case of shrinkage ratio, as shown in Figure 4, the neat PLA reveals shrinkage of length direction occur at over 5% which means the basic problem of PLA thermal stability. As above mentioned, the shrinkage of PLA composite was not observed by the flour in PLA composites role restriction of mobility of PLA molecular.

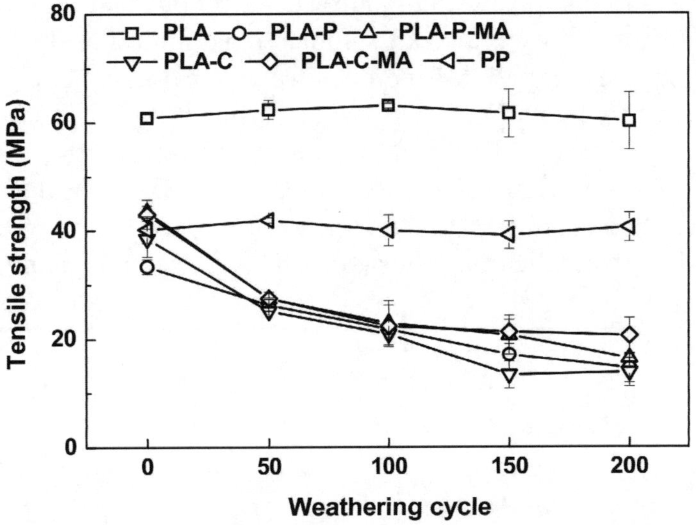

Figure 3. Tensile strength of PLA composites for thermal aging cycle.

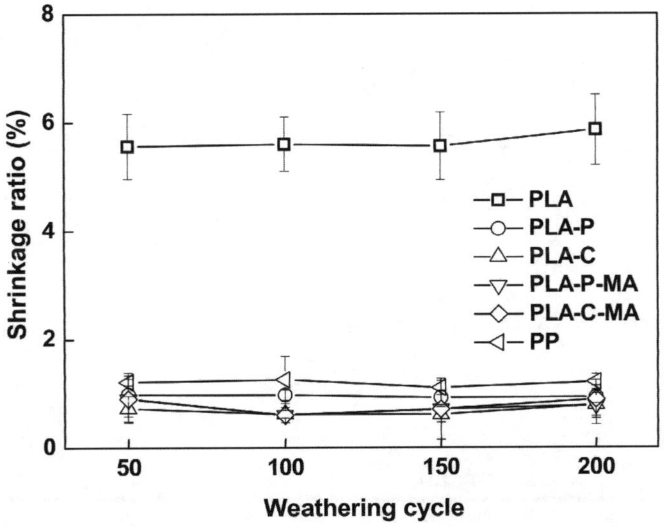

Figure 4. Shrinkage ratio of PLA composites for thermal aging cycle.

TVOC emission for bio-composites at isothermal condition

The inner temperature of car is changed according to four seasons, such as spring, summer, fall and winter in Korea. The maximum inside car temperature of spring and fall was known about 50□ but, it was the extreme temperature of car inside as 90□ in summer. So, we scheduled five temperatures from 30□ to 90□ to find optimal baking treatment condition for car interior material. The TVOC emission factors of the PLA and PLA bio-composites with different flours are shown on Figure 5. The TVOC of neat PLA was emitted from about 0.26 mg/m^2h to about 4.11 mg/m^2h with increasing temperature. And for both PLA bio-composite with pineapple flour (PLA-P) and with destarched cassava flour (PLA-C), the temperature increased from about 0.30 mg/m^2h to about 3.72 mg/m^2h and from about 0.19 mg/m^2h to about 8.74 mg/m^2h. TVOC emission factors of all samples increased gradually according to rising temperature. It may be VOCs in composites had more active mobility at higher temperature which was easy to emit out of composites. Below 70□, all samples have a similar TVOC emission factor. But, above 70□, both PLA-P composite and PLA-C composites which are higher TVOC emission factor than that of neat PLA, because of rapidly natural VOCs, such as furfural (2-furancarboyaldehyde), emission starting. Figure 6 shows of the TVOC of PBS and PBS bio-composites with different flours. Below 50□, all sample has a similar TVOC emission factor. But, TVOC of PBS is emitted rapidly from about 0.77 mg/m^2h to about 17.41 mg/m^2h at from 50□ to 90□. PBS is produced by the condensation reaction of the glycols 1,4-butanediol and aliphatic dicarboxylic acid, which is succunic acid [Fujimaki 1998].

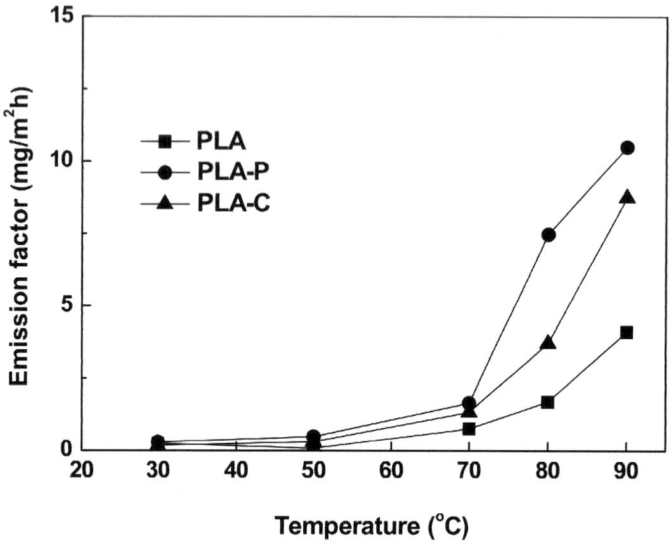

Figure 5. TVOC emission factor of PLA and PLA composites.

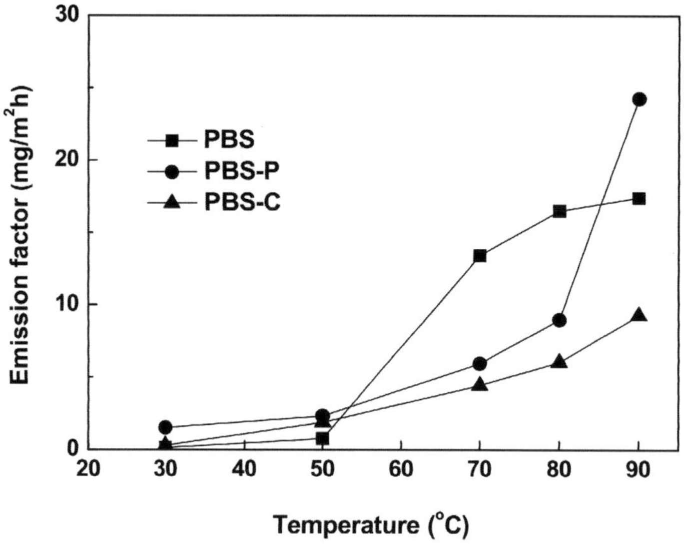

Figure 6. TVOC emission factor of PBS and PBS composites.

Conclusion

The HDT of PLA bio-composites blended with flour was higher than neat PLA and we also confirmed the increase of HDT according to MAPLA. According to thermal aging test, all PLA bio-composites decreased the tensile strength for wider gap between PLA matrix and flours for shorten molecular of PLA and 3wt% of MAPLA is not effect as compatibilizing agent in thermal aging test. However PLA and PP was not observed the strength decrease. The PLA was found about 6% shrinkage ratio from thermal aging test for low thermal stability. The TVOC of neat PLA was emitted from about 0.26 mg/m^2h to about 4.11 mg/m^2h with increasing temperature. And for both PLA bio-composite with pineapple flour and with destarched cassava flour, the temperature increased from about 0.30 mg/m^2h to about 3.72 mg/m^2h and from about 0.19 mg/m^2h to about 8.74 mg/m^2h. TVOC emission factors of all samples increased gradually according to rising temperature. Above 70℃, both PLA bio-composites were higher TVOC emission factor than that of neat PLA, because of rapidly furfural (2-furancarboyaldehyde) emission. PLA emitted 1, 4-Dioxane, known a carcinogenic substance. But, 1, 4-Dioxane could be by 30wt% added flour. Below 50℃, All sample has a similar TVOC emission factor. But, TVOC of PBS is emitted rapidly at from 50℃ to 90℃, because of succinic acid emission from pyrolysis of PBS.

References

1. Duek, EAR., CAC. Zavaglia, WD. Belangero (1999), "In vitro study of Poly(lactic acid) pin degradation," *Polymer*, 40, pp. 6465-6473.
2. Favaro, M., R, Mendichi, F. Ossola, U. Russo, S. Simon, P. Tomasin, P.A. Vigato (2006), "Evaluation of polymers for conservation treatments of outdoor exposed stone monuments. Part I: Photo-oxidative weathering," *Polymer Degradation and Stability*, 91, pp. 3083–3096.
3. Fujimaki T. (1998), "Processability and properties of aliphatic polyesters, 'BIONOLLE', synthesized by polycondensation reaction," *Polymer Degradation and Stability*, 59, pp. 209-214.
4. Huda, HS., L. T. Drzal, M. Misra, AK. Mohanty (2006), "Wood-Fiber-Reinforced Poly(lactic acid) Composites: Evaluation of the Physicomechanical and Morphological Properties", *Journal of Applied Polymer Science*, 102, pp. 4856-4869.
5. Weir N-A., F.J. Buchanan, J.F. Orr, D.F. Farrar, A. Boy (2004), "Processing, annealing and sterilisation of poly-l-lactide," *Biomaterials*, 25, pp. 3939–3949

The Effect of Mono- and Poly-Organic Acids on Thermoformed Starch-Protein Composites

Sudarsan Mukhopadhyay and Charles Onwulata
Dairy Processing and Products Research Unit, USDA-ARS, Eastern Regional, Research Center, Wyndmoor, PA 19038, USA

Abstract
Bioplastic materials derived from agri-proteins and starch are often difficult to process in high throughput plastic polymer molders due to their high moisture content and low melt properties. Starch and protein materials degrade above 140°C, exhibiting discoloration and low mechanical strength. In this study, we investigated the effects of the presence of acetic acid (AA), citric Acid (CA) and alkyl citrate analogs such as trimethyl citrate (TMC) and triethyl citrate (TEC) for improved cross linking and mechanical properties of thermoformed starch and protein biomaterial. A mixture of 40% whey protein isolate (WPI) and 30% corn starch in an aqueous solution of either AA, CA, or an alkyl citrate analog, in amounts ranging from 0 to 30 % (w/w) was used to form the bio-based composites. The components were mixed with roller blades in a torque rheometer for 45 min at the preset initial temperature of 70°C. At the completion of reactive mixing, thermoformed composite pastes were molded into ASTM D4065 strips in a Carver press for 30 min at 121°C. The dynamic mechanical properties were affected by high compounding moisture contents of 30 % (AA), 29 % (CA), 24% (TMC), 20 % (TEC) and 34 % (control). Flexibility increased with decreased moisture after molding with AA and CA or alkyl citrates.. Result trends indicate an increasing storage modulus of the molded materials with increasing concentration of organic acids. Further reduction in moisture increased elasticity. One composite, 15% TEC yielded better elastic modulus and mechanical property.

Introduction
World industrialization has generated substantial quantities of petroleum-based plastics over many years, which are not readily biodegradable and contributes to environmental pollution. This is an unsustainable situation. Hence, there is a growing demand for use of renewable agricultural sources, plant or animal origin, to develop eco- friendly biobased materials (Comstock et al.). Starch and cellulose are the most abundant and cheapest sources of biopolymers and widely used to create bioplastics (Greer, D. 2006). However, the use of starch as biomaterials, has some limitations. The main limitations are poor water resistance, oxidative instability, and variability in quality. Due to rapid degradation trend and hydrophilic characteristics, starches produce biomaterial with poor mechanical properties (Silva et al., 2006; Demirgoz et al., 2000). Proteins from animal such as whey protein or plant source such as soy (Lodha and Netravali, 2005), corn (Wu et al., 2003), cottonseed (Grevellec et al., 2001), and wheat (Cherian et al., 1995) have demonstrated several advantages for their desirable properties. Protein-based bioplastics have a high capacity for thermosetting modification because protein denaturation favors the development of a wide variety of materials. Although, the composites prepared from starch and protein such whey protein expected to have improved thermo-mechanical properties, the functionality of such biomaterials can be enhanced dramatically by thermal processing such as molding or extrusion, in the presence of additive or plasticizer. In the

absence of plasticizer, materials are brittle and difficult to handle (Gontard, et al., 1993). Plasticizer of small molecular sizes fits between the chains of polymer, spacing them apart, reducing intermolecular forces, and depress glass transition temperature of starch- proteins composites. Phthalic acid is widely used as additive for plastic making. Phthalates are used in preparation of a variety of plastic polymers, such as commercial cellulose ester based plastics. However, phthalates are currently the subject of considerable media, legislative and scientific scrutiny for its health threat. Citrates are green plasticizers and are safer to use due to their excellent biodegradability. Onwulata and coworkers, recently reported on a technology for making new and improved dairy based bioplastic (DBP) from whey protein isolate and corn starch material (Onwulata, 2005). In this work, we report the use of safe and eco-friendly citric acid and alkyl citrate analogs for preparation of starch-protein composites and investigate their influence on the composite physical, thermal and mechanical properties.

Materials
Materials used for making the bioplastic composites were purchased from different manufacturers as follows: whey protein isolates, Provon 190® (Glanbia Foods Inc., Richfield, ID); Corn starch (A.E. Staley Mfg. Co, Decatur, IL), Glacial acetic acid and citric acid (Mallinckrodt Baker, Inc., Phillipsburg, NJ), Trimethyl citrate (TCI America, Portland, OR) and Triethyl citrate (Spectrum, Trenton, NJ).

Methods

Torque Rheometry
The starch and protein-based formulations used were mixed in an Intelli-Torque Plasti-Corder torque rheometer (C.W. Brabender, South Hackensack, NJ) with a pair of roller blades using time and pre-mixer speed adjustments to obtain a homogeneous mixture with no clumps. The materials were mixed at 70 °C for 45 min at 60 rpm. The torque rheometer generated curves were then analyzed using flow behavior analysis software available in the Brabender Software WinMix version 3.2.16 (Duisburg, Germany). Compression molding of samples from torque rheometry was conducted using a Carver Laboratory Press Model B (Carver Instruments, Summit, NJ). The protein-starch mixture was placed in standard stainless steel plate (76.2 × 88.9 mm) with five slit opening for compressing rectangular test bar shapes (12.7 × 5 × 3 mm) used for thermo mechanical property analyses. The bars or materials were heated to 121.1°C, and held under constant pressure of 34.5 MPa for 30 min. The stain less steel plate was cooled to ambient immediately after completion of compression molding and the test bars were removed and was kept in a controlled humidity desiccator until analysis.

Moisture & Volatile Properties
Moisture content of the molded bio-plastics was determined by AOAC method no. 925.09 (AOCS, 2000), using a vacuum oven. The molded pieces were thinly sliced to provide more surface area for moisture and volatiles to escape more rapidly and provide more accurate reading.

Dynamic Mechanical Analysis (DMA)
The mechanical properties of the samples were analyzed using a dual cantilever beam multifrequency test. The width and thickness of each specimen was determined by an average of three measurements, recorded and used for the analysis. Each sample was mounted in a Q800 DMA (TA Instruments, New Castle, DE), cooled to -50°C, held isothermally for 5 min then the temperature was ramped to 175°C at 3°C/min for frequencies of 1 and 10 Hz. Tan Delta, Loss Modulus and Storage Modulus were analyzed at 1 Hz using TA Advantage software (TA Instruments, New Castle, DE).

Differential Scanning Calorimetry (DSC)
DSC and MDSC thermal properties of the plastic blends were measured on a Q100 DSC (TA Instruments, New Castle, DE). Samples were prepared by cutting them into small round pieces of approximately 7 – 15mg that fit into hermetically sealed aluminum DSC pans. The thermal profile of each specimen was determined using the DSC conditions: isothermal at 5°C for 5 min, then ramp to 200°C at 5°C/min.

Tensile and bending tests
A Sintech 1/G universal testing machine (MTS Systems, Eden Prairie, MN) was used for the tensile and bending tests. Tensile specimens were mounted vertically in grips and pulled apart at 5 mm/s until breakage. The peak load, peak stress, and strain, elongation, and modulus at break were determined. Bending tests were conducted by placing specimens horizontally on a flat hollow cylinder and pressing in the center with a solid cylinder 6 mm in diameter moving downward at 0.85 mm/s. The strain at peak load, break load, break stress, and elongation and strain at break were measured.

Results and discussion
The experiments were carried out in three concentrations 5%, 15% and 30% using AA, CA and alkyl citrates TEC and TMC. However, for comparison purpose results at 15% additive level is presented. The moisture content of finished composites is shown in Figure 1. The control sample contained much higher moisture compared to all trial composites with organic acids or alkyl citrate analogs. Both TEC and TMC composite have similar values (26.7 % and 23.7%). TMC contains about 12% less moisture than TEC composite. The control sample was found to have excessively high value of moisture compared to the citric acid ester composites; for example the control moisture content was 39% and 56% higher compared to TEC and TMC composites respectively. Lower moisture content signifies a desirable and flexible finished material.

The result obtained from DMA analysis in shear mode is expressed as storage modulus of the bioplastics (Figure 2). The addition of AA or CA lowered the modulus compared to control (133.1±21.2). The alkyl citrates, however, improved the quality of the composite as evidenced by higher modulus values in the presence of these plasticizers. The modulus increased, compared to control, by about 13% in presence of TMC and 27% in presence of TEC. On the basis of modulus values, TEC provides the most desirable elastic property among the polymer composites.

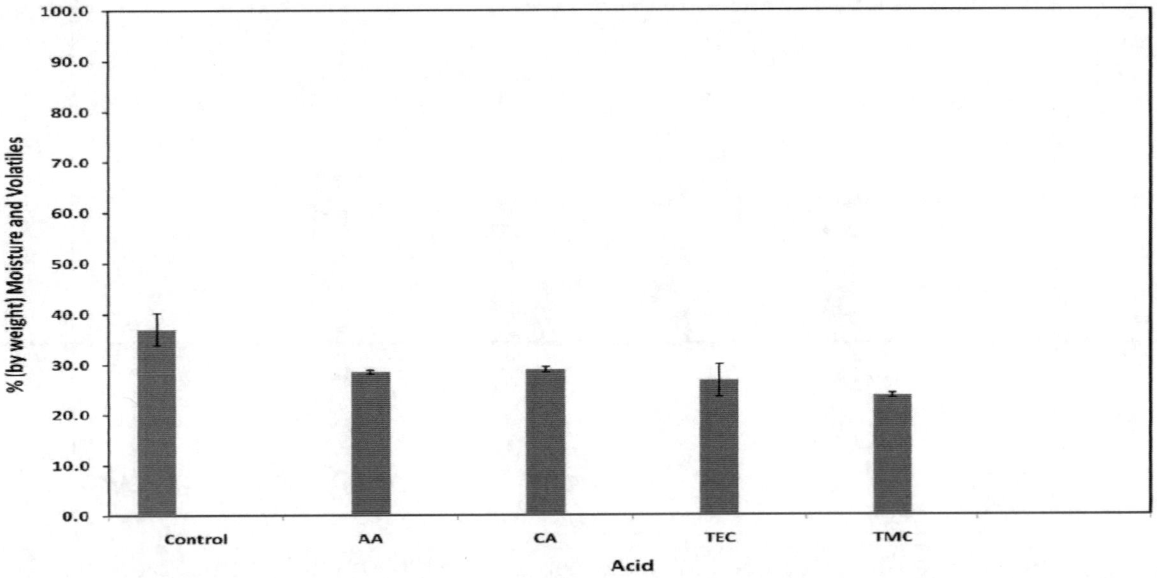

Figure 1. Moisture contents of bioplastic materials made with 15 % (w/w) concentration of various organic acids and acid esters

The melting point and glass transition temperature (T_g) of bioplastic composites are shown in Figure 3 and Figure 4, respectively. The addition of organic acids and alkyl citrates depressed the melting point of bioplastics by 11 % (alkyl citrate) to 18% (organic acids). In contrast to the melting point, the glass transition temperatures were somewhat time-temperature dependent for the biomaterials. The glass transition temperature of the control sample was 17.9 ±1.2 °C. The addition of TEC at a level of 15%, depressed the glass transition temperature by about 4 °C for the bio-composite. Although, for thermoplastics, T_g value may be expected to be reduced in the presence of plasticizer due to the embedment of plasticizer molecules between the starch-protein chain, the T_g value for TMC based composite actually increased significantly. More experiments are needed to explain this anomaly.

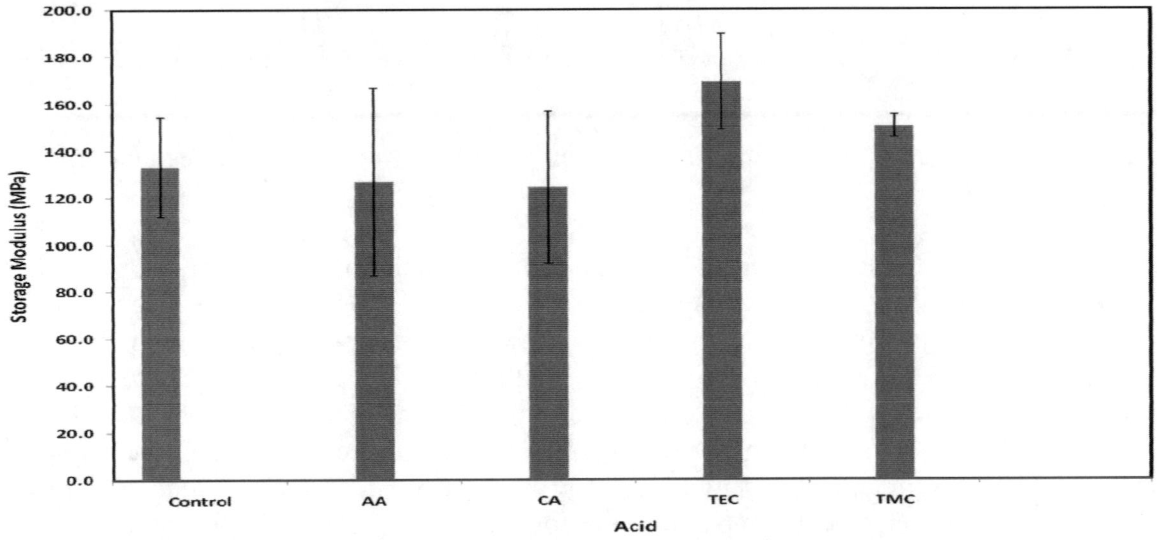

Figure 2. Influence of organic acids and acid esters (15% w/w) on the plastic property of starch–dairy protein based biomaterial

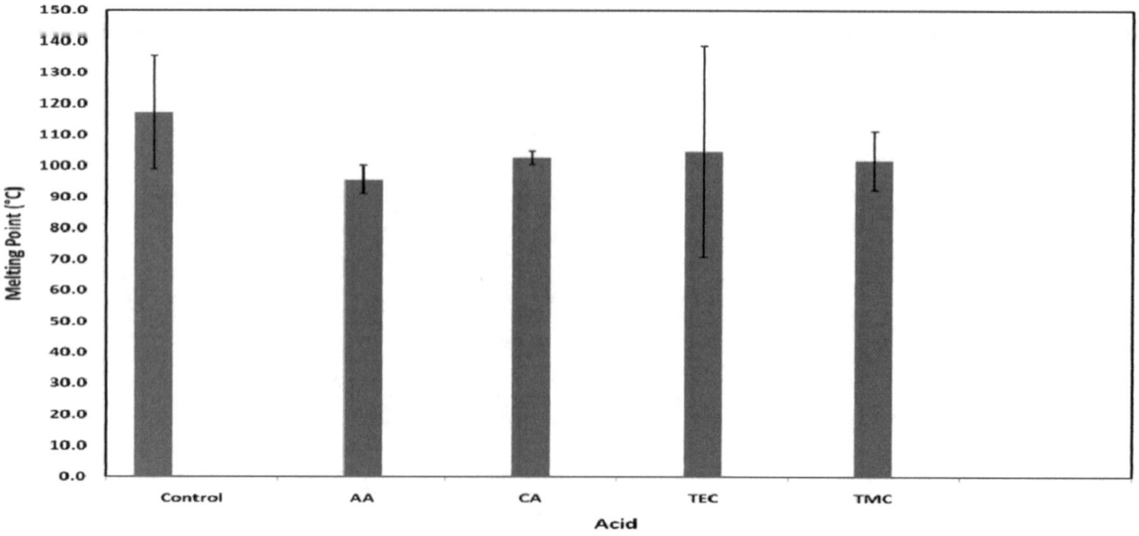

Figure 3 . Melting points for various bioplastic composites prepared with organic acids and alkyl citrates.

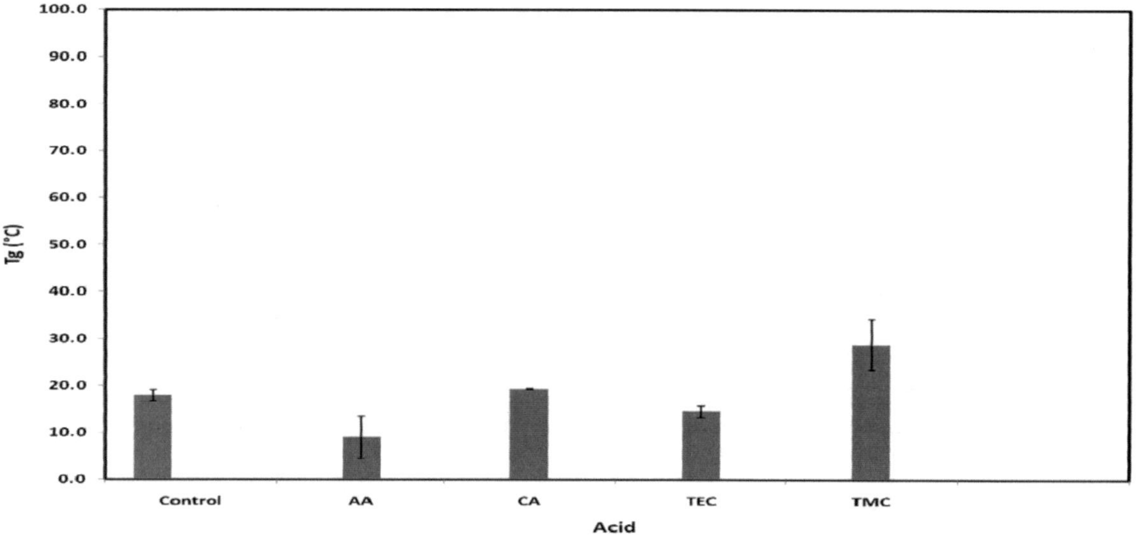

Figure 4 . Organic acid and alkyl citrates influence on glass transition temperatures of starch-whey protein composite in DSC.

The linear dynamic mechanical properties of starch-protein biomaterials in presence of organic acids and alkyl citrate plasticizers were measured from -50 °C to +175 °C over the frequency range of 1 to 10 Hz to develop correlations between the viscoelastic properties of composite materials and their impact and related deformation sensitivities. A typical DMA curve for TEC15 is given in Figure 5. It is revealed from the figure that the blend containing 15 % TEC has two distinct states. At lower temperatures (< 0 °C), the composite exhibited high storage modulus compared to high temperature ranges (> 0 °C) where the modulus was reduced rapidly. From the time lag in displacement compared to the applied force, the damping property of the material was estimated as shown in the figure by tan δ curve. A more realistic value for glass transition temperature for the composite was obtained by calculating the temperature corresponding to tan δ curve peak. The T_g values calculated from DMA curve for the composites made with organic acids and alkyl citrates were in good agreement with the DSC results.

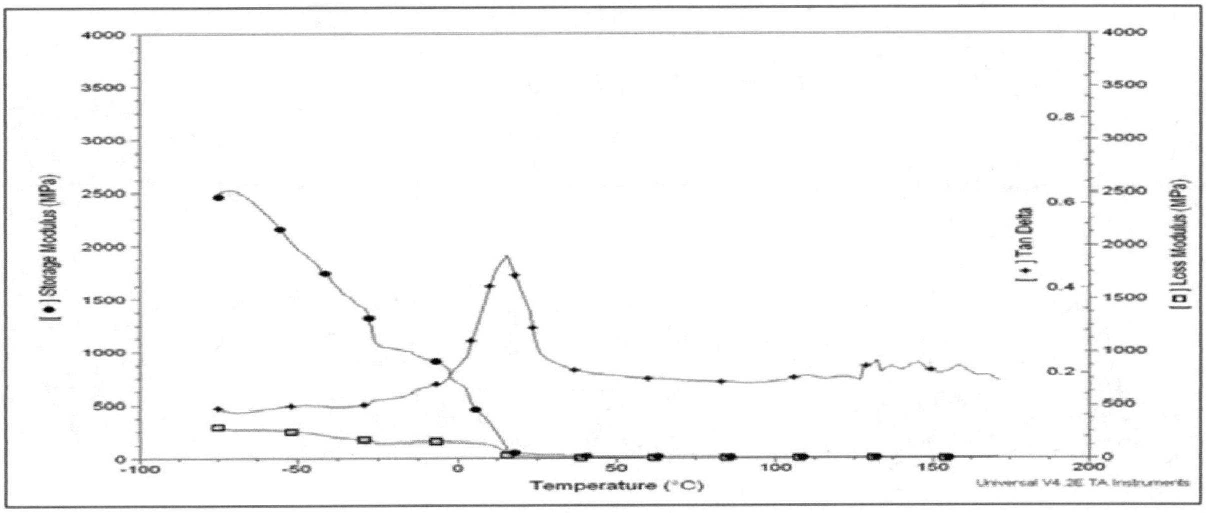

Figure 5. DMA curve for biocomposite with 15% TEC (w/w) in the matrix.

The tensile properties of the starch-whey protein plastic material determined using MTS system is given in Table 1. The properties of biomaterial made with alkyl citrates such as TEC and TMC provided better tensile strength compared to organic acids in the matrix (Table 1). The elongation of material with TMC (15%) was almost double that of the control sample. The TMC composites reflected best tensile strength.

Table 1. Influence of acids and alkyl acid citrates on composite tensile properties

Acids /Acid citrates, 15% (w/w)	Tensile-Peak Load, N	Elongation at Break, mm	Tensile-Modulus, MPa
Control	8.2±4.4	3±0.9	0.9±0.9
AA	16±7.2	8.1±3.9	1.4±0.9
CA	11.5±0.5	3.7±1.1	1.4±0.1
TEC	22.4±4	4.3±2.0	2.2±0.6
TMC	46.5±6.9	6.1±1.0	3.0±1.2

Conclusions

New starch-dairy protein based bioplastic was developed. Organic acids and alkyl esters behaved differently influencing the material property. TEC at 15 % level imparted the best viscoelastic property. However, in terms of tensile strength, TMC based composite material at 15% was the best. Initial screening indicated that both TMC and TEC are desirable plasticizers. Further experiments needed to be conducted with these additives to optimize their beneficial properties.

Acknowledgment
Authors like to thank Ms. Audrey Thomas, Mr. Eric Tilman and Dr. Michael Tunic for their help with this investigation.

References

AOAC. 2000. *Official Methods of Analysis*, 14th ed. Association of Official Analytical Chemists, Washington, DC.

Cherian, G., Gennadios, A., Weller, C., Chinachoti, P., 1995. Thermomechanical behavior of wheat gluten films: effect of sucrose, glycerin, and sorbitol. Cereal Chem. 72, 1–6.

Comstock, K., Farrell, D., Godwin, C., & Xi, Y. 2004. Website: http://depts.washington.edu/poeweb/students/gradprograms/envmgt/2004symposium/GreenPackagingReport.pdf

Gontard, N., Guilbert, S., Cuq, J.L., 1993. Water and glycerol as plasticizers affect mechanical and water vapor barrier properties of an edible wheat gluten film. J. Food Sci. 58, 206–211.

Grevellec, J., Marquie, C., Ferry, L., Crespy, A., Vialettes, V., 2001. Processability of cottonseed proteins into biodegradable materials. Biomacromolecules 2, 1104–1109.

Greer, D. *BioCycle* 47(5), 43 (2006).

Lodha, P., Netravali, A.N., 2005. Thermal and mechanical properties of environment-friendly 'green' plastics from stearic acid modified-soyprotein isolate. Ind. Crops Prod. 21, 49–64.

Onwulata, C. I.,. Liu, L. S., Finkenstadt, V., and Willett, J. L. Proceedings of the United States – Japan cooperative program in natural resources (UJNR), Honolulu, Hawaii, October 21-18, (2005).

Silva, M. C. , Ibezim, E. C., Ribeiro, T. A. A., Carvalho, C. W. P., and Andrade, C. T. Ind. Crops & Prod. Vol. 24, 46: (2006).

Wu, Q., Sakabe,, H. and Isobe, S., Ind. Eng. Chem. Res. 42, 6765 (2003).

Chain Scission Reaction in Hydrolytic Degradation of Polylactic Acid

Shri Ramaswamy, Department of Bioproducts and Biosystems Engineering, University of Minnesota, St. Paul, MN

Yuyang Liu, Department of Bioproducts and Biosystems Engineering, University of Minnesota, St. Paul, MN

Ulrike Tschirner, Department of Bioproducts and Biosystems Engineering, University of Minnesota, St. Paul, MN

The in vitro degradation behavior of polylactic acid (PLA) has been examined in terms of degree of degradation during an incubation period of up to 15 days. High performance liquid chromatograph and gel permeation chromatography (HPLC-GPC) were employed to characterize their degradation profiles. The degradation was carried out in a sealed system at 80°C. After degradation the amount of the degradation species were measured on HPLC and the molecular weight of degradation species were determined using GPC according to polystyrene standards and polylactic acid standards. The extent of random scission reaction was evaluated by the ratio of the number of degraded polymers to the number of repeating units in initial PLA chains, and extend of end scission reaction was evaluated by the ratio of the number of monomers degraded to the number of repeating units in initial PLA chains.

A significant decrease in molecular weight was observed in the first three days due to random scission reaction. After three days, the degradation is mainly governed by end scission reaction and the molecular weight drops slowly. The results confirmed hydrolysis of ester linkage as the primary mechanism of degradation of PLA. The gradient degradation and auto-catalysis affect of lactic acid in the degradation of PLA matrix were also studied. Although a great number of condensation polymers undergo random scission reaction, different hydrolytic degradation rates for internal and chain-end bonds have been reported. The different rates were attributed to differences in the electronic environment near the reaction sites. Efforts are underway in ascertaining which chain-end (carboxyl, hydroxyl or both) of PLA degrades faster than the internal bonds.

See more of this Session: Biobased Materials - I
See more of this Group/Topical: Forest and Plant Bioproducts Division

Surface Modification of Cellulose Fibers by Starch Grafting

Delong Song, School of Chemical and Biomolecular Engineering, Georgia Institute of Technology, Atlanta, GA

Chunxu Dong, IPST at Georgia Tech, Georgia Institute of Technology, Atlanta, GA

Yulin Deng, School of chemical and biomolecular engineering, Georgia Institute of Technology, Atlanta, GA

Grafting hydrogel polymer, such as polyacrylamide, acrylic acid, poly(n-propylacrylamide), etc, has been attracting more research interests. In this study, a technique that can easily graft raw starch on unrefined or slightly refined cellulose fibers is developed. The grafted starch functions similarly as wood fibrils generated by refining. The effects of grafting conditions, including pH, temperature, fiber consistency, crosslinker dosage, and starch dosage on grafting performance were investigated. The results indicate that grafted starch can significantly improve fiber-fiber bonding. Other advantages of by starch grafting will also be discussed.

See more of this Session: Biobased Materials - I
See more of this Group/Topical: Forest and Plant Bioproducts Division

Nanoenhanced Biomaterial

Manju Misra, School of Engineering and the Department of Plant Agriculture, University of Guelph, Guelph, ON, Canada

Progress

See more of this Session: Biobased Materials - II
See more of this Group/Topical: Forest and Plant Bioproducts Division

Flax Fiber Interfacial Bonding

Jonn Foulk, Cotton Quality Research Station, United States Department of Agriculture- Agricultural Research Service, Clemson, SC

Michael Fuqua, Mechanical Engineering Dept., North Dakota State Univ, Fargo, ND

Chad Ulven, Mechanical Engineering Dept., North Dakota State Univ., Fargo, ND

Mercedes M. Alcock, Composites Innovation Centre Manitoba Inc., Winnipeg, MB, Canada

Measured flax fiber physical and chemical properties potentially impact bonding and thus stress transfer between the matrix and fiber within composites. These first attempts at correlating flax fiber quality and biofiber composites contain the initial steps towards identifying key flax fiber characteristics that influence composite performance. Results demonstrate that fiber pullout and ILSS testing can both be considered viable methods for determining interfacial bonding strength for natural fibers. This study determined that among the fiber and matrix interfacial adhesion, fiber characteristics such as fiber elongation, wax content, bacterial content, fiber thickness, density, Ca and Mg content demonstrate a correlation to biofiber composite performance.

See more of this Session: Biobased Materials - II
See more of this Group/Topical: Forest and Plant Bioproducts Division

Flax Fiber Interfacial Bonding Improvements

Chad Ulven, Mechanical Engineering Department, North Dakota State University, Fargo, ND

Shanshan Huo, Mechanical Engineering Department, North Dakota State University, Fargo, ND

Michael Fuqua, Mechanical Engineering Department, North Dakota State University, Fargo, ND

Venkata S. Chevali, Mechanical Engineering Department, North Dakota State University, Fargo, ND

Flax fiber has become increasingly attractive as reinforcement in polymer matrix composites for structural applications. Due to its lighter weight, flax fiber reinforced composites can perform similar to glass fiber reinforced composites on a specific property comparison. However, in order maximize natural fiber reinforced composite performance; fiber surface modification and/or matrix manipulation are required. The interface between fiber and matrix plays an extremely important role in composite properties because it facilitates the load transfer between phases. Therefore, the key to improved performance is in controlling the chemical and mechanical properties of the fiber-matrix interface. Surface treatment is one way to reduce the hydrophilicity of the fiber by reacting the hydroxyl groups in the fibers in order to improve the adherence between flax fiber and polymer matrix. Matrix manipulation is another way to obtain a more congruent interaction between the flax fiber and polymer matrix. In this presentation, a multitude of different unidirectional flax fiber/vinyl ester composites were produced via a modified vacuum assisted resin transfer molding (VARTM) process. Higher specific properties (tensile, flexure, interlaminar shear, fiber bundle pull-out, etc.) of chemically treated flax fiber reinforced composites over those of glass fiber reinforced composites were obtained and will be discussed.

See more of this Session: Biobased Materials - II
See more of this Group/Topical: Forest and Plant Bioproducts Division

Renewable Macroporous Nanocomposites Via Emulsion Templating

Nadine Graeber, Chemical Engineering, Imperial College London, London, United Kingdom

Angelika Menner, Chemical Engineering, Imperial College London, London, United Kingdom

Alexander Bismarck, Chemical Engineering, Imperial College London, London, United Kingdom

Open porous polymer foams have been synthesised via the polymerisation of high internal phase emulsions (HIPE). HIPEs are defined as emulsions having an internal phase volume of at least 74 %. Commonly, HIPEs consists of relatively uniform, polyhedral, aqueous droplets, separated by thin films of continuous monomer phase. The polymerisation of the continuous organic phase of HIPEs leads to open porous polymer foams so called polyHIPEs. PolyHIPEs gained increasing interest due to their unique properties such as low density, high porosity and high degree of interconnectivity. PolyHIPEs offer exciting opportunities for membrane applications in harsh environments as alternative to hollow fibre membrane modules and even in tissue engineering, i.e. as scaffolds. PolyHIPEs have also been explored for use as absorbents, demulsifiers and catalyst supports in solid phase chemistry. Here we report on the synthesis of nanocomposites foams based on renewable resources. We used acrylated soy bean oil as main component of the continuous monomer phase of the emulsion template. Furthermore, modified bacteria cellulose was added to either phase as nanoscale reinforcement of the resulting polymer foams. We are able to produce more environmentally friendly systems which might be suitable replacements of commonly used petrochemical based polymer foams.

See more of this Session: Biobased Materials - II
See more of this Group/Topical: Forest and Plant Bioproducts Division

Highly Aligned Cellulose Nanowhisker Reinforced PVA Nanofiber Webs Via Electronspinning Method

Jihoon Lee, School of Chemical and Biomolecular Engineering, Georgia Institute of Technology, Atlanta, GA

Yulin Deng, School of chemical and biomolecular engineering, Georgia Institute of Technology, Atlanta, GA

Cellulose nanowhiskers reinforced poly(vinyl alcohol) (PVA) nanofiber web were fabricated using electrospinning technique. The electrospun fiber diameter is ~300nm. Electrospun fiber alignment was achieved by using a rotating collector. A degree of alignment was measured by SEM observation and analyzed with an computer enhanced image analysis program. Tensile strength and modulus of cellulose nanowhiskers reinforced nanocomposites web was improved compared to pure PVA thin films. Mechanical properties of random distributed electrospun fiber thin films and aligned fiber thin films were compared. Relations of reinforcement effects, fiber alignment and cellulose nanowhiskers alignment have been investigated. Glutaraldehyde (GA) and Polyaminoamide-epichlorohydrin (PAE) were introduced to cellulose nanowhiskers - PVA electrospun fibers as cross-linkers. Cross-linking effects on mechanical properties of electrospun fibers have been studied. Detailed cross-linking mechanism of PAE has been investigated.

See more of this Session: Biobased Materials - II
See more of this Group/Topical: Forest and Plant Bioproducts Division

Hydrothermal Carbonization of Switchgrass for Biochar Production

Sandeep Kumar, Chemical Engineering, Auburn University, Auburn, AL

Lingzhao Kong, Chemical Engineering, Auburn University, Auburn, AL

Ram B. Gupta, Chemical Engineering, Auburn University, Auburn, AL

Hydrothermal treatment, an environmental friendly technology, provides a novel method for the utilization of biomass. The hydrothermal carbonization (HTC) process, a traditional but recently revived method, presents superior characteristics that make it a promising route of wide potential application for converting biomass to biochar. In this work, switchgrass, a major energy crop, was processed by HTC method in a batch reactor. The effect of temperature (200-230°C), reaction time (1-4 hrs), and pressure (14-276 bar) are examined. The obtained biochar was characterized for molecular structure by infra-red spectroscopy (FTIR) and scanning electron microscopy (SEM) techniques. The heating value of biochar were analyzed using a bomb calorimeter and results were compared with the theoretical heating value determined based on elemental analysis. With the increased severity, the lignocellulosic structure of switchgrass is broken and comparatively less fibrous biochar having a high energy density is obtained. The FTIR and SEM analyses show that biochar contained rich new functions groups. The results demonstrate that the HTC process can provide a promising route for carbonaceous functional materials and higher energy density fuel for use in a variety of applications.

See more of this Session: Biobased Materials - III
See more of this Group/Topical: Forest and Plant Bioproducts Division

Inulin Surfactants for Drug Delivery

Ezequiel Delgado, Guillermo Toriz and Julio C. Arboleda. Universidad de Guadalajara. Guadalajara, Jal., Mexico.

Introduction

Fructans are reserve polysaccharides found in plants such as onion, garlic, artichoke, chicory, and other unusual sources such as the agave plant [1, 2, 3]. The fructan known as inulin is the better studied and characterized in terms of its chemical and physical properties. Inulin is a linear polydisperse polysaccharide of 40-60 repeat units of fructose, which are linked by $\beta(2\rightarrow1)$ bonds and usually has one glucose moiety at the reducing end. The unique type of linkages makes inulin, and all other fructans, a non-digestible polysaccharide in the upper digestive tract that is only metabolized in the colon.

The colon has been considered by many to be an innocuous organ that has simple functions of absorbing water and electrolyte and the formation, storage and expulsion of faecal material. However, the colon suffers from several diseases that include ulcerative colitis, inflammatory bowel disease, Crohn's disease and colorectal cancer. Therefore, it is very desirable to target drug delivery to the colon, since it would ensure direct treatment at the disease site, lower dosing and fewer systemic side effects. On targeting of drugs to the colon it is important to control the rate and/or site of drug release from oral formulations for the purposes of improving patient compliance and treatment efficacy [4].

Owing to its macromolecular nature and the lack of specificity in binding with most cells, inulin has been used as a marker for studying the intracellular uptake of drug delivery systems [5]. However inulin could be used as drug delivery agent itself since owing its chemical nature, fructans can only be metabolized in the human colon by special kinds of bacteria, such as the *Bifidobacteria* genre. In fact, B*ifidobacteria* are known to promote health in the colon by taking over disease-causing bacteria such as *Vibrio cholera, Escherichia coli*, *Shigella*, *Listeria,* etc. The property of passing virtually intact through the upper intestinal tract makes inulin an ideal compound to deliver drugs specifically to the colon and allowing the design of such systems.

As the majority of polysaccharides, inulin has three hydroxyl groups available at each fructose unit, which give access to additional chemical functionality or more precise delivery. Production of surfactants from inulin can be achieved by esterification reactions of the hydroxyl groups with long aliphatic chain acid chlorides.

To emulsify different drugs, surfactants have to hold specific relations between their hydrophobic and hydrophilic components (HLB values). Therefore the synthesis of inulin derivatives has to be optimized in order to obtain the required HLB value of the particular drug that is going to be encapsulated. This can be achieved by adjusting the molar relationship between inulin and the hydrophobic reactant

In this work we aimed to produce inulin based surfactants in order to prepare emulsions with vitamin E. Esterification of chicory inulin with stearoyl chloride was carried out at different molar ratios to control the HLB of the surfactant. Once the emulsions were prepared micelles were crosslinked with glutaraldehyde and dried. The rationale is to form structured microcapsules where the drug is confined in the hydrophobic tails of the surfactant whereas the polar polysaccharide is directed toward the aqueous medium. In this way the exterior core could be metabolized in the colon, allowing the aperture of the capsule and thereafter the release of the drug.

Materials and Methods

Materials

Chicory inulin (fructose:glucose ratio 44:1) and stearoyl chloride (both form Sigma-Aldrich) and N-Methyl-2-pyrrolidone, NMP, (Sigma) were used without further purification. All the other reagents (acetone, hexane, polysorbate 20, sorbitan monolaurate, liquid Vaseline and E vitamin) were purchased from local suppliers (Analityka, Fermont, Chemico Especialidades químicas, Oxiteno, Drogas la Paz, and Botica Nueva) and used also as received.

Methods

HLB Determination: HLB was determined according to Griffin [6].
Esterification reactions: Inulin (22g dry basis 122.116 mmol) was dissolved in 90 g of NMP at room temperature. This solution was distilled in high vacuum to a concentration of 15% w/w to remove moisture in the system. The esterification reaction was carried out at 30 °C under a nitrogen blanket for 24 h. The molar relation stearoyl chloride:inulin was varied in order to obtain different HLB. The resulting solution was mixed with water (35 g), filtrated and washed four times with acetone in order to remove NMP and other impurities. The estearification product paste was dried with an air stream at room temperature.

The molecular weight of chicory inulin was determined by GPC. Inulin esters were dissolved in THF and analyze by SEC at a flow of 1 mL/min and 300 psi pressure. ATR-FTIR spectra were obtained in a Perkin-Elmer Spectrum GX at 4 cm-1 resolution and 32 scans.

Vitamin E emulsions. The required amounts of vitamin E, ater and inulin esters were placed in a test tube and heated at 50 C for 5 min. The mix was then stirred in a Ultra Turrax equipment at 20450 rpm for 10 min, keeping the temperature constant. The emulsion thus prepared was cooled under a stream of fresh water.

Results

It was possible the synthesis of micro spheres from stearoyl-inulin esters from chicory without the use of additional surfactants for emulsification. It was observed that the relation between the degree of esterification and the HLB of the surfactant is linear, similar to what is known for surfactants of simpler structures. An agreement between the HLB of the surfactant and the substance to be encapsulated did not guarantee a stable emulsion. The best emulsion was obtained with a surfactant with an HLB of 8 and a 0.86 molar relation of stearoyl:inulin.

Surprisingly, the surfactant prepared with an HLB of 8 did not yield stable emulsions, in spite of matching the HLB of the vitamin E. The HLB scale seems to provide only a guide to predict the ability of surfactants derived from long chain carbohydrates.

Under the conditions of the esterification reaction studied in this work, caramelization of the inulin took place. Degradation reactions produced darkening of the products and reduced the efficiency of the reactants. There was evidence of an increase in the molecular weight of the surfactant affecting its performance as emulsifier. In order to avoid caramelization of sugars and degradation products it was necessary to eliminate water from the starting material, reduce the temperature and use a nitrogen atmosphere.

The rate of release of vitamin E from microspheres prepared from stearoyl-inulin esters emulsions was dependant of the degree of crosslinking achieved by reaction with glutaraldehyde.

In order for this approach to produce microspheres from inulin surfactants for the encapsulation of drugs to the colon, a more rigorous purification procedure should be employed after the synthesis. The presence of NMP solvent traces, stearic acid or stearoyl chloride left unreated and fractions of fructans that do not participate in the reaction affected the emulsification power of the surfactants.

It was possible the synthesis of micro spheres from stearoyl-inulin esters from chicory without the use of additional surfactants for emulsification. It was observed that the relation between the degree of esterification and the HLB of the surfactant is linear, similar to what is known for surfactants of simpler structures. An agreement between the HLB of the surfactant and the substance to be encapsulated did not guarantee a stable emulsion. The best emulsion was obtained with a surfactant with an HLB of 8 and a 0.86 molar relation of stearoyl:inulin. Surprisingly, the surfactant prepared with an HLB of 8 did not yield stable emulsions, in spite of matching the HLB of the vitamin E. The HLB scale seems to provide only a guide to predict the ability of surfactants derived from long chain carbohydrates.

Under the conditions of the esterification reaction studied in this work, caramelization of the inulin took place. Degradation reactions produced darkening of the products and reduced the efficiency of the reactants. There was evidence of an increase in the molecular weight of the surfactant affecting its performance as emulsifier. In order to avoid caramelization of sugars and degradation products it was necessary to eliminate water from the starting material, reduce the temperature and use a nitrogen atmosphere.
The rate of release of vitamin E from microspheres prepared from stearoyl-inulin esters emulsions was dependant of the degree of crosslinking achieved by reaction with glutaraldehyde.

In order for this approach to produce microspheres from inulin surfactants for the encapsulation of drugs to the colon, a more rigorous purification procedure should be employed after the synthesis. The presence of NMP solvent traces, stearic acid or stearoyl chloride left unreated and fractions of fructans that do not participate in the reaction affected the emulsification power of the surfactants.

References

1 R. Starbird, V. Zúñiga, E. Delgado, B. Saake and G. Toriz, Journal of Biobased Materials and Bioenergy 1, 238 (2007).
2 C. V. Stevens, A. Meriggi and K. Booten, Biomacromolecules 2, 1 (2001).
3 G. Toriz, E. Delgado and V. Zúñiga, e-Gnosis on line 5, 1 (2007).
4 A. Basit and J. Bloor, in B U S I N E S S B R I E F I N G : P H A R M A T E C H (Touch Briefings, 2 0 0 3), p. 185.
5 H. Essien, J. Y. Lai and K. J. Hwang, Journal of Medicinal Chemistry 31, 898 (1988).
6 W. C. Griffin, Journal of the Society of Cosmetic Chemists 1, 311 (1949).

Lignocellulosic-Fiber-Reinforced Thermoplastic Composites: Effect of Processing-Induced Stresses On Mechanical Properties

Venkata S. Chevali, Mechanical Engineering and Applied Mechanics, North Dakota State University, Fargo, ND

Michael A. Fuqua, Mechanical Engineering and Applied Mechanics, North Dakota State University, Fargo, ND

Shanshan Huo, Mechanical Engineering and Applied Mechanics, North Dakota State University, Fargo, ND

Chad A. Ulven, Mechanical Engineering and Applied Mechanics, North Dakota State University, Fargo, ND

Natural fiber reinforced thermoplastic composites that use distillery or agricultural co-products such as distiller's dried grains with solubles (DDGS) or sunflower hull (SFH) as reinforcements have emerged as substitutes to glass fiber reinforced composites. These composites exhibit uncompromised mechanical properties at a fraction of the material and processing costs of synthetically reinforced systems. A low-cost commodity thermoplastic matrix enables the incorporation of a wide range of lignocellulosic fibers, including short particulate-type fibers and long bast-type fibers.

In the present investigation, the lignocellulosic fibers, i.e., DDGS and SFH, obtained from vendors are subjected to an appropriate mechanical size-reduction to enable their incorporation into a thermoplastic matrix by melt blending. Following this operation, the fiber surface is treated with a compatibilizing agent, such as maleic anhydride-grafted polypropylene in the case of polypropylene, or graft-polymerization with acrylic acid in the case of poly(methyl methacrylate). The fibers are sequentially compounded with the thermoplastic polymer in a co-rotating twin-screw extruder, water-quenched and pelletized to a required size, and injection molded.

As the polymer/reinforcement undergoes high shear in both the extruder and injection molding under temperatures reaching or exceeding 200 °C, the initiation of thermal and oxidative degradation of the fibers can occur, often indicated by discoloration and odor of the resulting polymer blend. In addition to degradation in strength, the decomposition of the fiber constituents, i.e., cellulose, hemicellulose, lignin, and starch, may lead to an additional decrease in fiber/matrix interfacial strength. This deterioration can result in reduction of the overall composite strength due to the inability of the weakened fibers to transfer load through the matrix. Thermal stability, a function of the extent of thermal and oxidative degradation, is critical and varies with fiber constituents, processing stresses, and processing durations.

In this study, the effect of processing on the mechanical properties is parametrically analyzed. The overall thermal degradation is studied as a function of extruder screw speed and subsequent material dwell time. Static mechanical properties in tension and flexure are correlated with thermal analysis using dynamic mechanical analysis, and molecular structure using Fourier-transform infrared spectroscopy. Scanning electron microscopy is used for fractography and for studying the surface morphology.

See more of this Session: Biobased Materials - III
See more of this Group/Topical: Forest and Plant Bioproducts Division

EXTRUSION FOAMING OF CELLULOSE FIBER REINFORCED POLYPROPYLENE COMPOSITES

T. Kuboki[1], R. Koyama[1], J.W.S. Lee[1], C. B. Park[1], J. Zhang[2], Q. Wu[2], and M. Sain[3]

1. Microcellular Plastics Manufacturing Laboratory
Department of Mechanical and Industrial Engineering
University of Toronto, Toronto, ON, Canada
2. School of Mechanical and Automotive Engineering
South China University of Technology, Guangzhou, China
3. Centre for Biocomposites and Biomaterials Processing
Faculty of Forestry, University of Toronto, Toronto, ON, Canada

Abstract

This paper investigates the effect of polypropylene type and cellulose content on the foaming behaviour of cellulose fiber reinforced polypropylene composites in extrusion. Two types of polypropylene (linear and branched structures) are used as a polymer matrix. The thermal properties of the composites are characterized by a differential scanning calorimeter (DSC), and the viscosity of the composites is evaluated by a rotational rheometer. The foaming behaviour of the composites is examined using an extrusion foaming system, in which carbon dioxide (CO_2) is used as a physical blowing agent for foams. The results suggest that with the increase of cellulose content, the cell density increases, whereas the void fraction decreases, but only up to a cellulose content of 20 wt%. When the cellulose content is increased to 40 wt%, the void fraction then increases. The results also indicate that the two types of polypropylene have a minimal effect on the foaming behaviours of the cellulose fiber composites.

Introduction

Polypropylene (PP), which is one of the most widely-used plastics in the world, offers high strength, heat and chemical resistance, non-hygroscopicity, and relatively low price. Due to the excellent properties of solid PP, there is a strong demand for PP foams, in particular, from the automotive, insulation, and food and protective packaging industries. However, the PP foam market is much smaller than other plastic foam markets, such as polyurethane, polystyrene, and polyvinyl chloride, because in comparison, the quality of existing PP foams is not satisfactory. Standard PPs with a linear molecular structure lack the required extensional rheological properties in the melt phase for the production of low density foams with a fine and controlled cell structure.

In this context, a long chain, branched, high melt strength (HMS) PP, which combines both HMS and extensibility in the melt phase, has been developed in order to improve foamability [1-5]. In practice, it is reported that the HMS PP foams exhibit a higher expansion ratio than linear PP foams [6].

It is known that foaming decreases the strength and modulus of plastics. Therefore, it is necessary to reinforce the plastic foams in order to utilize them for structural applications. Glass fiber is often used as reinforcement for PPs. However, environmental and safety concerns have compelled researchers and engineers to seek alternative reinforcements.

Cellulose fiber could be an option for the reinforcement of PPs. It possesses superior mechanical and thermal properties. It has low density, a high aspect ratio, and nonabrasiveness. Furthermore, since cellulose fiber contains little extractives, the emission of volatiles can be reduced during processing, which may provide cellulose composite foams with better cell morphology than other natural fillers or fiber composite foams.

In this paper, cellulose fibers with different contents are blended with two types of PPs (linear and branched), and the effects of the type of PPs and cellulose content on the foaming behaviors of the cellulose/PP composites are examined in the extrusion foaming process with CO_2.

Experimental

Materials

Two types of PP (HF136MO and WF420HMS, Borealis) were used as polymer matrixes. Both PPs are similar in melt flow index (20 and 22 g/10 min (230 °C/2.16 kg) for HF136MO and WF420HMS, respectively) and density (0.908 and 0.905 g/cm^3, respectively, for HF136MO and WF420HMS), but differ in polymer chain structure. One has a linear structure, and the other, a branched structure (HF136MO and WF420HMS are named linear and branched, respectively, hereafter). Cellulose fiber (TC150, CreaFill Fibers Corp.) was employed as the reinforcement material. The fiber is 1-2 μm in thickness, 20 μm in width, and 120 μm in length. Maleated polypropylene (PP-MA) (Polybond 3150, Chemtura) was used to modify the interface between the fiber and the polymer matrix. The cellulose fiber was dried in an oven at 105 °C for 12 hours before compounding. The PP, cellulose fiber, and PP-MA were compounded using a twin-screw extruder (Micro-27, Leistritz). In this study, 8 types of composite materials are prepared. The formulations of the composites are summarized in Table 1. The cellulose fiber contents vary from a range of 0 to 40 wt%, but the PP-MA content is fixed at 3 wt% for all of the composite materials in order to examine the effect of the fiber content on the cell morphology.

Characterization of Thermal Properties

The melting and crystallization temperatures, and crystallinity of the samples were measured using a differential scanning calorimetry (DSC, TA Q2000) with a thermal analysis data station. Each sample was heated to 200 °C at a rate of 10 °C/min. The samples were then allowed to cool down at a rate of 10 °C/min. A second heating scan was carried out at a rate of 10 °C/min. Thermal properties, such as melting temperature (T_m), crystallization temperature (T_c), and heat of fusion (ΔH_{exp}), were determined from the DSC thermograms. The degree of crystallinity (χ_c) of the PP is evaluated using Eq. (1):

$$x_c = \frac{\Delta \mathrm{H}_{\mathrm{exp}}}{\Delta \mathrm{H}^*} \times \frac{1}{\mathrm{W}_{\mathrm{f}}} \qquad (1)$$

where ΔH_{exp} is the experimental heat of fusion determined from DSC; ΔH^* is the heat of fusion of the fully crystalline PP (210 J/g); and W_f is the weight fraction of PP in the composites.

Characterization of Viscosity

The samples were subjected to oscillatory shear in a dynamic stress rheometer (TA ARES) with parallel plates at a gap of 1.0 mm. In all cases, the dynamic stress sweeps were initially performed to define the limits of the linear viscoelastic regime. The dynamic frequency sweep experiments were then conducted to measure the complex viscosity (η^*) over a frequency range of 0.1-100 rad/s at 180 °C. All measurements were performed under nitrogen to prevent polymer degradation or moisture absorption.

Extrusion Foaming

The PP, PP/PP-MA blends, or cellulose/PP pellets were processed in a single-screw, three-zone extruder with a gas injection system. In this study, carbon dioxide (CO_2) was used as an environmentally friendly physical blowing agent. The CO_2 content for the polymer matrix in the composite was fixed at 3 wt%, i.e., 0, 2.4, and 1.8 wt% for the composites with 0, 20, and 40 wt% fiber, respectively. The temperature in this extruder was maintained at 180 °C. A static mixer, heat exchanger, and filament die (L/D: 1.2"/0.06" for polymers and 0.6"/0.06" for composites) were attached to the downstream of the barrel. Their temperatures were set at 180 °C. The die temperature was reduced gradually from 180 °C to 145 °C in order to prevent gas loss in the extrudate which would consequently reduce the density. The foam samples were collected at each stabilized processing condition and characterized for foam density, void fraction, cell morphology, and cell density.

The foam density was first measured using the water displacement method (ASTM D792-00). The expansion ratio (Φ) was then calculated on the basis of the ratio of the bulk density of PP, PP/PP-MA, or cellulose/PP composites (ρ_o) and the measured density of the foam sample (ρ_f). The void fraction was subsequently determined using Eq. (2):

$$\text{Void Fraction } (\%) = \left(1 - \frac{1}{\Phi}\right) \times 100 \tag{2}$$

In order to observe cell morphology, the extruded foam samples were immersed into liquid nitrogen and fractured in a brittle manner. The fractured surface was coated with a thin layer of platinum using a sputter coater (SC7620, Quorum Technologies) and the microstructure was examined using a scanning electron microscope (SEM) (JSM-6060, JEOL). The cell densities are calculated from the SEM micrographs using Eq. (3):

$$\text{Cell Density} = \left(\frac{nM^2}{A}\right)^{\frac{3}{2}} \times \Phi \tag{3}$$

where n is the number of bubbles in the micrograph; A and M are the area and magnification factor of the micrograph, respectively; and Φ represents the expansion ratio.

Results and Discussion

Thermal Properties and Viscosity

Figure 1 summarizes the thermal properties of the PPs, PP/PP-MA, and cellulose composites (Figure 1(a) for melting temperature, Figure 1(b) for crystallization temperature, and Figure 1(c) for crystallinity). Figure 1(a) shows that for all the given cellulose contents, linear and branched PP composites have similar melting temperatures, and the melting temperatures of both composites are minimally affected by the cellulose contents.

On the other hand, the branched PP presents a higher crystallization temperature than the linear PP (Figure 1(b)), which is consistent with the results reported from previous studies [7-9]. The branching of PP increases the crystal nuclei density, promotes faster crystallization and thus, a higher crystallization temperature. It is also noted that the addition of PP-MA increases the crystallization temperature slightly for both linear and branched PPs. When cellulose fibers are incorporated in PP/PP-MA, the crystallization temperature of the linear PP composites is increased with the increase of the cellulose content while that of branched PP composites changes minimally. As a result, the linear and branched PP composites possess similar crystallization temperatures at 40 wt% cellulose content. It is speculated that the cellulose fibers act as a nucleating agent in the linear PP matrix during the nucleation stage, resulting in an increase in the crystallization temperature with increasing cellulose content. A similar phenomenon is reported for other natural fiber or filler composites [10-12]. The reason why the addition of cellulose has minimal influence on the crystallization temperature of the branched PP needs to be further studied.

In Figure 1(c), linear PP composites show slightly higher crystallinity than the branched PP composites. The figure also shows that the addition of cellulose decreases the crystallinity of the PP matrix for both the linear and branched PP composites. This is because the cellulose fibers interfere with the crystallization during the crystal growth, which results in a decrease in the overall crystallinity level of the PPs with increasing fiber content.

Figure 2 illustrates the frequency dependence of η^* (Figure 2(a) for linear PP composites and Figure 2 (b) for branched PP composites). The linear and branched composites show similar values of shear viscosity at a given cellulose content in the all frequency range. For both composites, shear-thinning behavior increases with the increase of cellulose content.

Foaming Behavior

Two dies were used for the foam extrusion system in order to keep die pressures higher than solubility pressures in the extrusion process. The dies used were the same in diameter (0.06") but different in length (1.2" for the pure PP and PP/PP-MA blends and 0.6" for the cellulose composites). Figure 3 charts the die pressure in the extrusion process of the samples with 3 wt% CO_2 for the PP matrix at different die temperatures (Figure 3(a) from the pure PP and PP/PP-MA blends and Figure 3(b) from the cellulose composites). Both figures indicate that the die pressures at which the foam samples were taken (more than 1300 psi) are high enough to dissolve the injected CO_2 into the PP matrix [13]. Therefore, a single CO_2/PP phase is formed at the die inlet. For all the samples, the die pressures increase as the die temperatures are decreased or cellulose content is increased because the viscosity of the molten polymers and composites is increased at the die.

Figure 4 depicts the foam density with respect to different die temperatures. For the 0 wt% cellulose composites (Figure 4(a)), the foam density of all the samples decreases as the die temperature is decreased. The order of the density is as follows for all the die temperatures: branched PP/PP-MA > linear PP/PP-MA > linear PP > branched PP. The results indicate that the addition of PP-MA increases the foam densities of the linear and branched PPs, although the density of PP-MA (0.91 g/cm^3) is very similar to that of PP matrixes (0.908 g/cm^3 for the linear PP and 0.905 g/cm^3 for the branched PP).

When cellulose fibers are incorporated with PPs (20 wt% cellulose for Figure 4(b) and 40 wt% cellulose for Figure 4(c)), the foam density of all the cellulose composites does not change much with a value of 0.8 g/cm^3 in relation to the die temperature, which means that the PP type and cellulose content have a minimal effect on the foam density when cellulose is added.

Figure 5 shows the cell density with respect to cellulose content. When PP-MA is applied to the linear and branched PPs, the cell density is increased by one order of magnitude. The cell density is increased further with an increase of cellulose content because the cellulose acts as a nucleating agent. The addition of cellulose increases the nucleation rate of bubbles in keeping with the principles of heterogeneous nucleation [14]. In heterogeneous nucleation, the activation energy barrier to nucleation is reduced in the presence of cellulose. As a result, the addition of cellulose to PPs increases cell nucleation. In addition to the enhanced cell nucleation, it is surmised that the addition of cellulose suppresses the cell growth and/or cell coalescence in the foaming processes. A higher fraction of polymer chains is confined by cellulose, and thus their mobility is restricted, which increases the melt viscosity of the composites (see Figure 2). Guo et al. [15,16] studied the effect of wood fiber content on the cell morphology of wood fiber plastic composites in extrusion foaming. They reported that a small amount of wood fiber improves cell morphology, but a large amount deteriorates the cell structure, probably due to the larger amount of volatiles and irregular distribution of the wood fiber. However, in this study, such optimum cellulose fiber content for cell density is not observed. This may be because cellulose fiber contains little extractives, which have reduced the emission of volatiles in the process. It should also be noted that the cell density is minimally affected by the difference in the polymer matrix for either linear or branched PPs, in all the ranges of cellulose content.

Figure 6 compares the void fraction of linear and branched composites with respect to cellulose content at a die temperature of 165 °C. For both composites, the void fraction is decreased when the cellulose content increases from 0 to 20 wt%, but increases as the cellulose content increases from 20 wt% to 40 wt%. This concavo curve of void fraction with cellulose content could be explained by the competition between the polymer matrix content in the composite and cell density. Up to 20 wt% cellulose, the decrease of polymer matrix content (from 100 wt% to 80 wt%) in a composite where foaming can take place, dominantly affects the void fraction, which decreases the void fraction of the composites. However, with an increase of 20 wt% to 40 wt% cellulose, it is the cell density that primarily influences the void fraction. It is reported that the void fraction is increased with the increase of cell density for high-density foams [17]. Therefore, it is speculated that the void fraction of cellulose composites is increased by the increase of cell density (see Figure 5).

Conclusions

The foaming behavior of linear and branched PP/cellulose composites is examined in extrusion. Pure branched PP show lower foam density than that of pure linear PP. However, the addition of PP-MA and cellulose suppress the superior expandability of pure branched PP, resulting in similar foam densities of linear and branched composites. On the other hand, the addition of PP-MA and cellulose increases the cell density of both linear and branched composites, and the cell density is increased with the increase of cellulose content. However, there is no significant difference in cell density between pure linear and branched PPs or their composites. For all the given cellulose contents, linear and branched composites demonstrate similar void fractions and show minimal values at 20 wt% cellulose.

Acknowledgement

This work was supported financially by AUTO21, the Consortium for Cellular and Microcellular Plastics (CCMCP), and the Natural Sciences and Engineering Research Council of Canada (NSERC). We also wish to express our thanks to Borealis, CreaFill Fibers Corp. and Chemtura for supplying materials.

References

1. M.B. Bradley, and E.M. Phillips, *SPE ANTEC Tech. Papers*, **36**, 717 (1990).
2. J.J. Park, L. Katz, N.G. Gaylord, *U.S. Patent*, 5,116,881 (1992).
3. U. Panzer, *Specialty polyolefins SPO 98* (1998).
4. M. Ratzsch, U. Panzer, A. Hesse, H. Bucka, *SPE ANTEC Tech. Papers*, **45**, 2071 (1999).
5. P. Park, *Proc. Foamplas 2000*, 35 (2000).
6. H.E. Naguib, C.B. Park, U. Panzer, and N. Reichelt, *Polym. Eng. Sci.*, **42**, 1481 (2002).
7. A.J. De Nicola, A.F. Galombos, and M.D. Wolcowicz, *Polym. Mater. Sci. Eng.*, **67**, 106 (1992).
8. X. Wang, C. Tzoganakis, and G.L. Rempel, *J. Appl. Polym. Sci.*, **61**, 1395 (1996).
9. H.E. Naguib, C.B. Park, and S.W. Song, *Ind. Eng. Chem. Res.*, **44**, 6685 (2005).
10. X. Colom, J. Canvate, P. Pages, J. Saurina, and F. Carrasco, *J. Rein. Plast. Comp.*, **19**, 818 (2000).
11. P. Ganan, and I., Mondragon, *J. Therm. Analys. Calori.*, **73**, 783 (2003).
12. N. Sombatsompop, A. Kositchaiyong, and E. Wimolmala, *J. Appl. Polym. Sci.*, **102**, 1896 (2006).
13. Y. Sato, K. Fujiwara, T. Takikawa, Sumarno, and S. Takishima, *Fluid Phase Equilibria*, **162**, 261 (1999).
14. J.S. Colton, and N.P. Suh, *Polym. Eng. Sci.,* **27**, 500 (1987).
15. G. Guo, K.H. Wang, G.M. Rizvi, Y.S. Kim, and C.B. Park, *PPS-20*, Paper #105, Akron, Ohio, June 20-24, 2004.
16. G. Guo, G.M. Rizvi, C.B. Park, and Y.H. Lee, *PPS-21*, Paper #3761, Leipzig, Germany, June 19-23, 2005.
17. J.W.S. Lee and C.B. Park, *Macromol. Mater. Eng.*, **291**, 1233 (2006).

Table 1. Material Information of the Composite Materials Prepared in This Study.

No	Linear PP (wt%)	Branched PP (wt%)	PP-MA (wt%)	Cellulose (wt%)
1	100		0	0
2	97		3	0
3	77		3	20
4	57		3	40
5		100	0	0
6		97	3	0
7		77	3	20
8		57	3	40

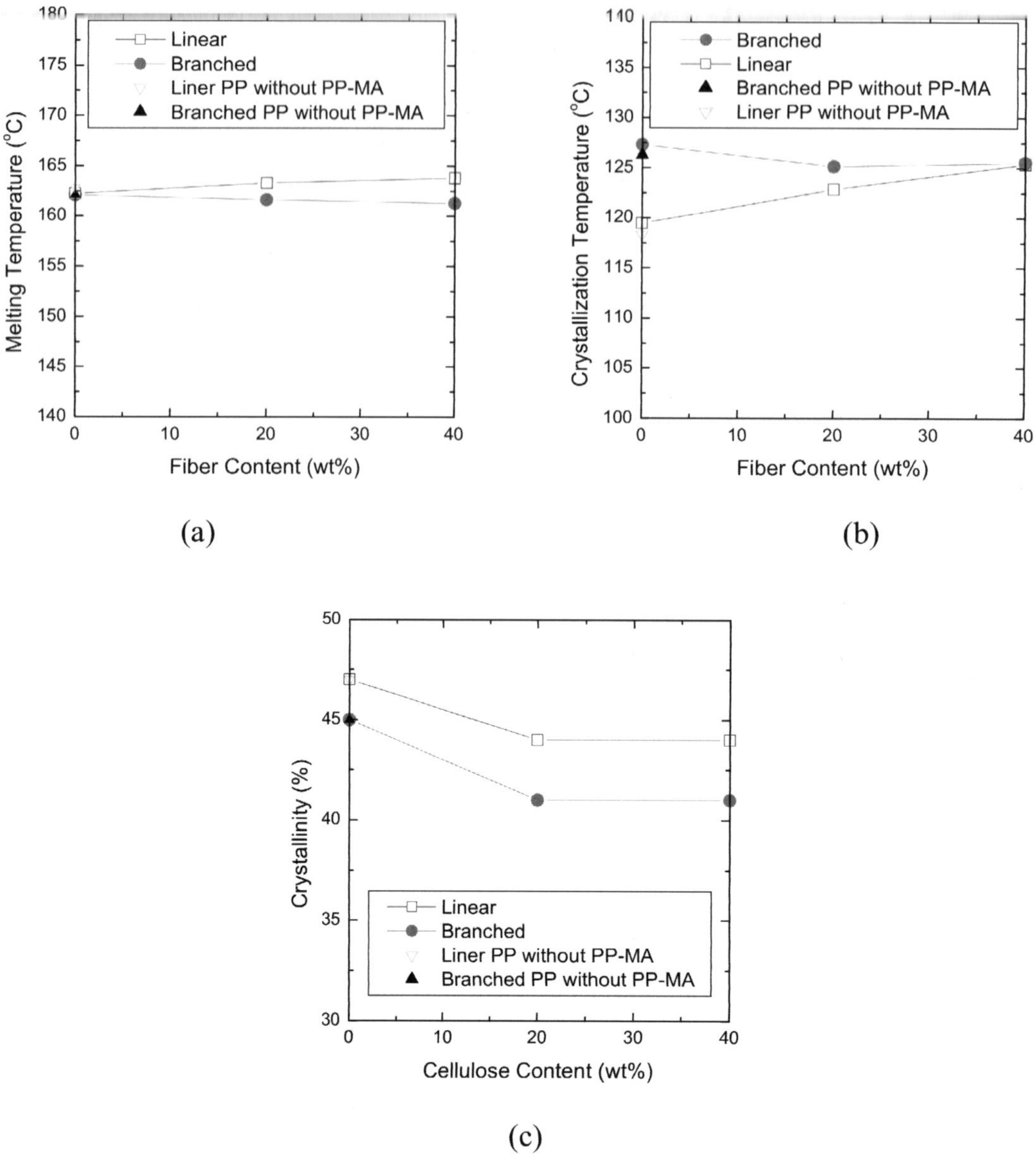

Figure 1. Thermal Properties of Cellulose/PP Composites as a Function of Cellulose Content: (a) Melting Temperature, (b) Crystallization Temperature, and (c) Crystallinity

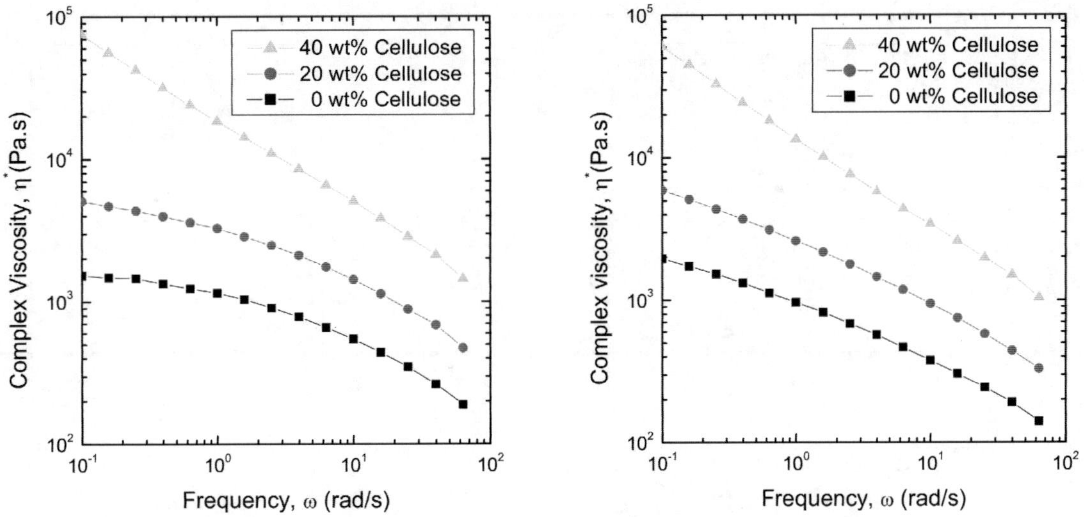

(a) (b)

Figure 2. Complex Viscosity (η*) of Cellulose/PP Composites as a Function of Frequency: (a) Linear PP Composites and (b) Branched PP Composites

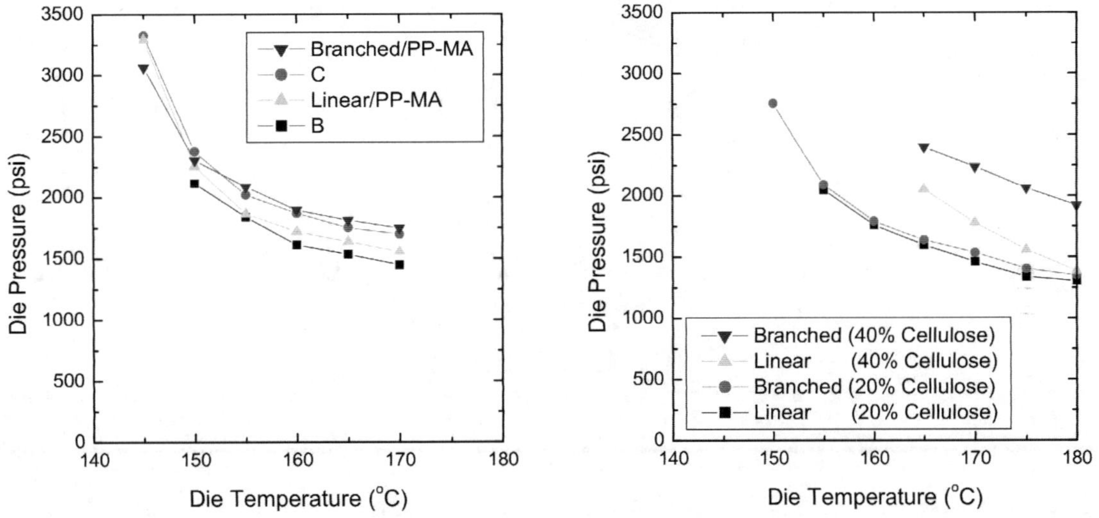

(a) (b)

Figure 3. Die Pressure as a Function of Die Temperature: (a) PPs and PP/PP-MA blends using 1.2"/0.06" Die and (b) Cellulose Composites using 0.6"/0.06" Die

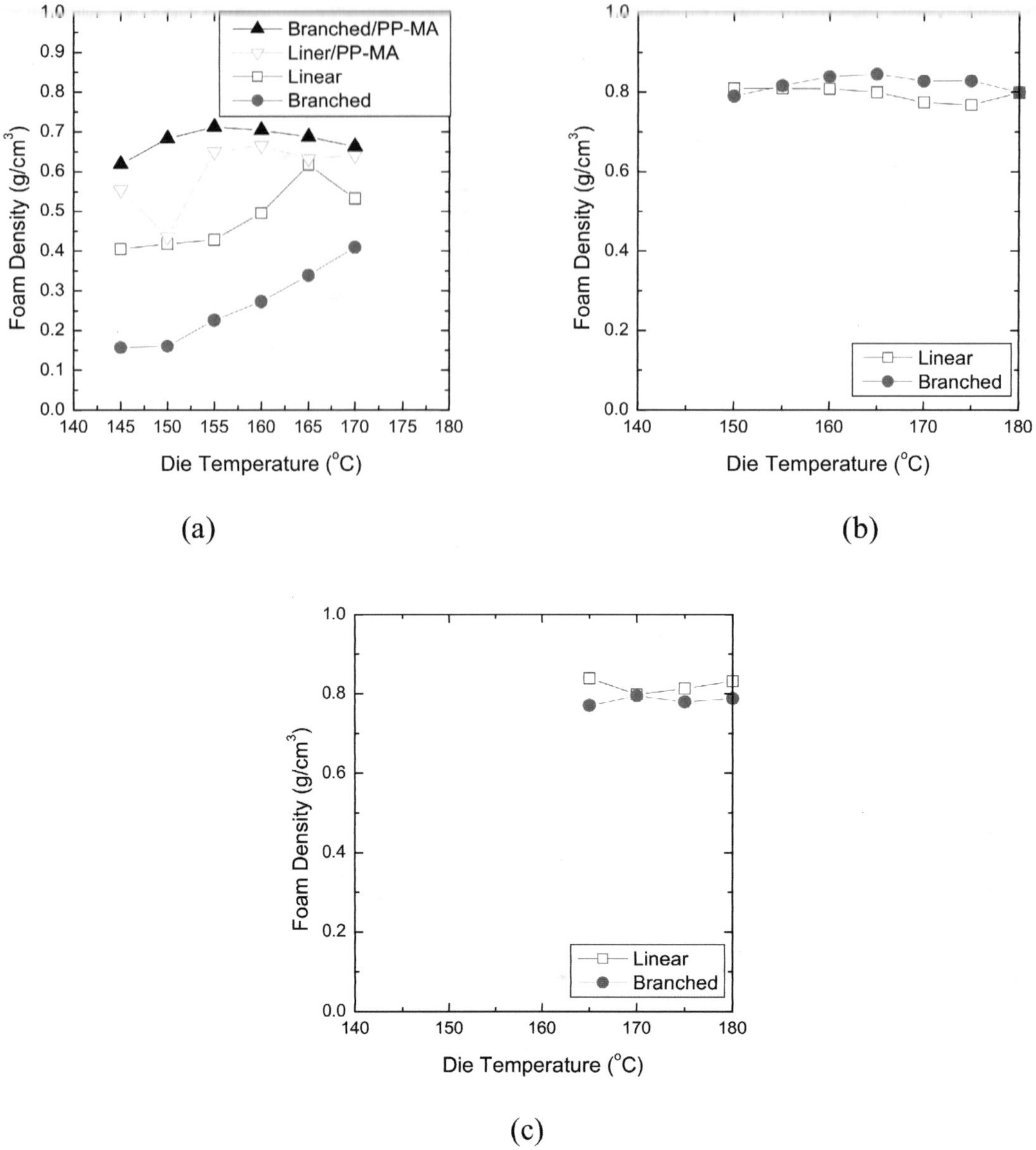

Figure 4. Foam Density of Cellulose/PP Composites as a Function of Die Temperature: (a) 0 wt% Cellulose, (b) 20 wt% Cellulose, and (c) 40 wt% Cellulose

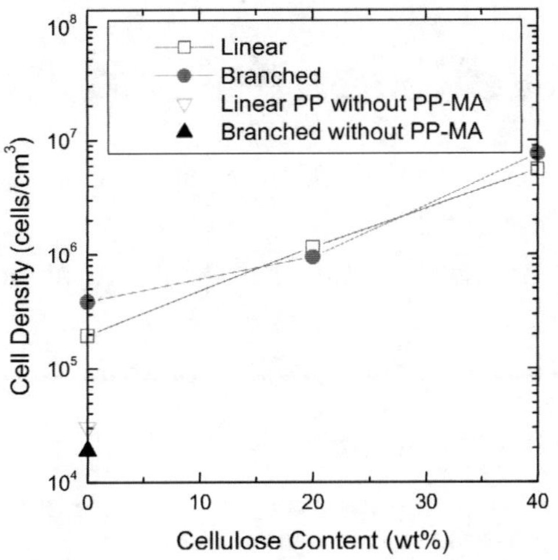

Figure 5. Cell Density of Cellulose/PP Composites as a Function of Cellulose Content at a Die Temperature of 165 °C

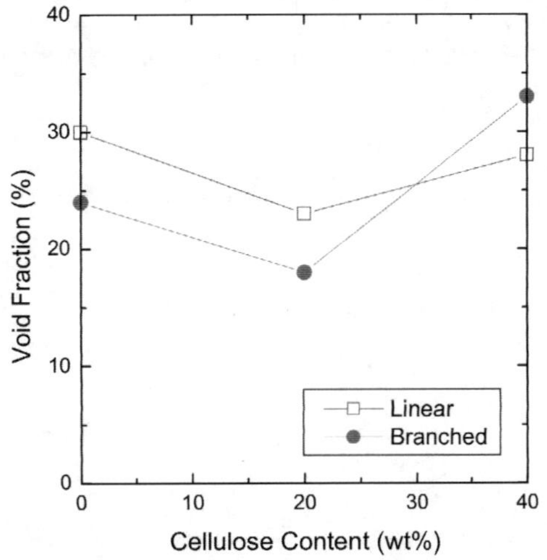

Figure 6. Void Fraction of Cellulose/PP Composites as a Function of Cellulose Content at a Die Temperature of 165 °C

Lignin Based Novel Blends and Composites

S. Sahoo, Bioproducts Discovery and Development Centre, Department of Plant Agriculture, Crop Science Building,, University of Guelph, Guelph, ON, Canada

Manju Misra, School of Engineering and the Department of Plant Agriculture, University of Guelph, Guelph, ON, Canada

Amar K. Mohanty, Department of Plant Agriculture & School of Engineering, Guelph, ON, Canada

Lignin is a byproduct/co-product produced in a huge amount from paper industries and bioethanol industries. Currently, Lignin is mainly used as fuel for boilers with a very negligible percentage utilized for chemical recovery, agriculture and value added material applications. In the present study, Green composites were fabricated from engineered lignin, switch grass fiber and a biopolyester using twin screw extruder followed by injection moulding. 1% newly discovered compatibilizer with active functionality was used to improve the performance of composites. Different formulations i.e. with varying lignin content (30%, 50%, and 65%), biomass-lignin hybridization (50% filler content) and compatibiliized composites were made to evaluate the desired material properties for specific applications. The specimens were characterized through dynamic mechanical analysis (DMA), tensile and flexural tests. Incorporation of lignin improved the tensile, flexural, thermo mechanical properties of composites. Biomass-lignin hybridization with 1 wt% compatibilizer improved the flexural strength by 165%, storage modulus by 480% and heat deflection temperature (97.5 0C) by 26 0C. Compatibilized hybrid composite achieved flexural strength of 75.2MPa.

See more of this Session: Biobased Materials - III
See more of this Group/Topical: Forest and Plant Bioproducts Division

Barriers and Status in Woody Biomass Pretreatment for Ethanol Production: Technology and Energy Cost

J.Y. Zhu[1,2*] and X.J. Pan [2]

[1] U.S. Forest Service, Forest Products Laboratory, Madison, WI

[2] Department of Biological Systems Engineering, University of Wisconsin, Madison, WI

ABSTRACT

About 370 million tons of woody biomass, accounting for 30% of the total biomass, can be sustainably produced annually in the United States (Perlack et al., 2005). To promote biodiversity, sustainable and healthy forest and ecosystem management, and to meet local and regional bioenergy needs, woody biomass will be a critical integral part of biomass supply mix in the future bio-economy. The major differences between woody and nonwoody (agricultural) biomass are their physical properties and chemical compositions. Wood is a structural material. Woody biomass is physically larger, structurally stronger and denser than agricultural biomass. Chemically, it has higher lignin and cellulose contents than agricultural biomass. As a result, woody biomass is more recalcitrant to microbial and enzymatic destructions than nonwoody biomass. This is particularly true for softwood species. More energy is required to overcome the woody biomass recalcitrance, especially energy consumed for wood size-reduction to the level for effective enzymatic processing. On the other hand, woody biomass has many unique features. For example, woody biomass can be harvested year round, which eliminates long-term storage. Its high density significantly reduces transportation cost. Furthermore, its high lignin content increases energy density. Its near-zero ash content eliminates dead load in transportation and processing. Moreover, woody biomass, in particular softwood species, generally has a lower content of pentoses than agricultural biomass, which is favorable for bioconversion to ethanol.

Substantial research efforts have been made in biochemical conversion of agricultural biomass for cellulosic ethanol production. Progress from these efforts can also be applied to woody biomass, especially those associated with downstream processing. However, the

---

jzhu@fs.fed.us

upstream processing such as pretreatment and saccharification for woody biomass are uniquely different from those for nonwoody materials due to the differences in their physical and chemical properties between the two feedstocks as outlined above. Particular attentions need to be paid to: (1) the effectiveness of pretreatment for complete wood cellulose saccharification, and (2) the energy consumption for woody biomass pretreatment, in particular for wood size-reduction to the level for effective enzymatic saccharification. Unfortunately, limited research efforts and resources have been devoted to these critical areas. This is also true in the United States despite cellulosic ethanol has been selected as a major alternative fuel for the next 15 years, i.e., 16 billion gallons cellulosic ethanol by 2022 (US Energy Security and Independence Act of 2007), and U.S. has 30% of the annually available biomass being woody biomass.

The present paper provided a comprehensive discussion on the key technical issues in woody biomass pretreatment, i.e., barriers to efficient cellulose saccharification, pretreatment energy consumption, in particular energy consumed for wood size-reduction, and criteria to evaluate the performance of a pretreatment. Net energy production is the ultimate goal of any biofuel technologies. To significantly reduce energy consumption in wood size-reduction, we proposed a new approach of **post-chemical pretreatment size-reduction** (Zhu et al., 2009a), instead of conducting *size-reduction prior to chemical pretreatment.* The later approach of conducting size-reduction before chemical pretreatment has been described in many expert reviews (Lynd, 1996; Yang and Wyman, 2008) and DOE Bioenergy Roadmap (DOE, 2005). Despite many disadvantages such as energy intensive, it has been widely practiced in the literature such as the recent comprehensive study on poplar wood bioconversion to ethanol (Wyman et al. 2009), conducted by the Biomass Refining Consortium for Applied Fundamentals and Innovation (CAFI) sponsored by the US Department of Energy. The **post-chemical pretreatment size-reduction** approach proves very effective for efficient enzymatic saccharification and reducing energy in size-reduction and has many benefits especially for woody biomass biochemical conversion (Zhu et al., 2009a). In-plant processing energy consumption for woody biomass can be reduced to the level for agricultural biomass by using this approach.

When evaluating and comparing the performance of pretreatment technologies for biomass bioconversion for cellulosic ethanol production, energy input should be an important factor to consider. The optimization of a pretreatment process should not be only simply based on total sugar production, but also on total energy consumption. We defined a concept of **pretreatment process energy efficiency** (kg/MJ), the total sugar recovery (kg/kg wood) divided by the energy consumption in pretreatment (MJ/kg wood). It was then used to evaluate the performances of three most promising pretreatment technologies, steam explosion (Soderstrom et al., 2004), organosolv (Pan et al., 2008), and sulfite pretreatment to overcome lignocelluloses recalcitrance (SPORL, Zhu et al., 2009b) for softwood pretreatment. From the limited literature data, the SPORL process appears to be the best process for woody biomass conversion especially for softwood species, in terms of total sugar production, process energy efficiency, lignin co-product potentials, feedstock versatility, and commercial scalability.

REFERENCES

DOE, 2005. Breaking the biological barriers to cellulosic ethanol: A joint research agenda. A research road map resulting from the biomass to biofuel workshop sponsored by The Dept of Energy. December 7-9, Rockville, MD.

Lynd, L.R., 1996. Overview and evaluation of fuel ethanol from cellulosic biomass: technology, economics, the environment, and policy. Annual Review of Energy and the Environment 21(1), 403-465.

Pan, X.J., Xie, D., Yu, R.W., Saddler, J.N., 2008. The bioconversion of mountain pine beetle-killed lodgepole pine to fuel ethanol using the organosolv process. Biotechnology and Bioengineering 101, 39-48.

Perlack, R.D., Wright, L.L., Turhollow, A., Graham, R.L., Stokes, B., Erbach, D.C., 2005. Biomass as feedstock for A Bioenergy and Bioproducts Industry: The Technical Feasibility of A Billion-Ton Annual Supply, Oar Ridge National Laboratory, ORNL/TM-2005/66, US Dept. of Energy, Oak Ridge.

Soderstrom, J., Galbe, M., and Zacchi, G., 2004. Effects of washing on yield in one- and two-step steam pretreatment of softwood for production of ethanol. Biotechnol. Prog. 20, 744-749.

Wyman, C.E., Dale, B.E., Elander, R.T., Holtzapple, M., Ladisch, M.R., Lee, Y.Y., Mitchinson, C., Saddler, J.N., 2009. Comparative sugar recovery and fermentation data following pretreatment of poplar wood by leading technologies. Biotechnol. Prog. 25, 333-339.

Yang, B., Wyman, C.E., 2008. Pretreatment: the key to unlocking low-cost cellulosic ethanol. Biofpr 2, 26-40

Zhu, W., Zhu., J.Y., Gleisner, R., Pan, X.J., 2009a. On energy consumption for size-reduction and enzymatic hydrolysis of Lodgepole pine, the 31st Symposium on Biotechnology for Fuels and Chemicals, May 3-6, San Francisco, also submitted to Bioresource Technology.

Zhu, J.Y., Wang, G.S., Pan, X.J., Gleisner, R., 2009b. Sulfite pretreatment for robust enzymatic saccharification of spruce and red pine. Bioresource Technology 100, 2411-2418

Kinetic Characterization of Xylose Monomer and Oligomer Concentrations During Dilute Acid Pretreatment of Lignocellulosic Biomass From Forests

Juan E. Morinelly, Chemical Engineering, Michigan Technological University, Houghton, MI

Jill R. Jensen, Chemical Engineering, Michigan Technological University, Houghton, MI

Matthew Browne, Chemical Engineering, Michigan Technological University, Houghton, MI

Tomas B. Co, Chemical Engineering, Michigan Technological University, Houghton, MI

David R. Shonnard, Chemical Engineering, Michigan Technological University, Houghton, MI

Dilute acid pretreatment achieves the saccharification of the hemicellulose portion of lignocellulosic biomass in a biomass to ethanol process. The yield of fermentable, monomer sugars during pretreatment is a crucial factor for the maximization of the overall process efficiency.

The kinetics of xylose formation and degradation has been investigated in more detail than that of any other sugar present in hemicellulose. This is the case given that xylose is often the main component with percentages of total hemicellulose by dry weight ranging from 27 to 91% depending on the type of feedstock [1]. The simplest kinetic mechanisms that has been proposed for this reaction is a two-step pseudo-first order irreversible reaction with Arrhenius-type kinetic constants [2]:

$$Hemicellulose\,(XH) \xrightarrow{k_1} Xylose\,(X) \xrightarrow{k_2} Degradation\ products\,(D)$$

However, experimental observations suggest that other models that include oligomeric intermediates and parallel reactions of slow and fast reacting hemicellulose phases (biphasic) could describe the reaction more accurately [3]. There is not a consensus on whether the biphasic behavior of hemicellulose is real or simply an artifact of experimental methods. On the other hand, oligomer concentrations have been measured [4-6] indicating that an oligomer intermediate should be included in kinetic modeling:

$$Hemicellulose\,(XH) \xrightarrow{k_1} Oligomer\,(XO) \xrightarrow{k_2} Xylose\,(X) \xrightarrow{k_3} Furfural\,(F) \xrightarrow{k_4} Degradation\ products\,(D)$$

The kinetics of xylose monomer and oligomer sugars under dilute acid hydrolysis for aspen, balsam, and switchgrass were investigated at various temperatures, acid concentrations, and reaction times using small-scale isothermal tubular reactors. The experimental data were fitted to a four-step kinetic model with first-order irreversible rate constants at each step. The maximum yield of xylose monomer for aspen and balsam show limited variability with changing reaction severity with maximum levels around 85% and 67% of initial xylan respectively, while switchgrass monomer levels varied significantly with very low yields observed at 0.25% wt. acid (30% of initial xylan) and relatively high yields at 0.75% wt. acid (80% of initial xylan). The measured furfural levels at the point where xylose monomer is maximized, relative to initial xylan, range from 1% to 6% for aspen and balsam, and 3% to 15% for switchgrass. Excellent agreement, both quantitatively and qualitatively, is shown between the experimental monomer data and its model fit. Although the maximum xylose oligomer concentrations as well as the initial points during oligomer formation are accurately predicted, the model with irreversible kinetic constants does not describe the oligomer data accurately at long times. The model predicts furfural levels comparable to those measured; however, the model overpredicts early data points and underpredicts late data points. Arrhenius kinetic parameters were derived for all three species at each step of the proposed model, and these parameters were of the same order as prior literature values. Preliminary results of the effect of aspen solids loading on reaction kinetics show that a 50% decrease in reactor solids loading (10% to 5%) results in faster rates of reaction, particularly for the degradation of xylose monomer which increases by a factor of almost 3.

References

1. McMillan, J.D., *Process for Pretreating Lignocellulosic Biomass: A Review. NREL/TP-421-4978.* 1992.

2. Saeman, J.F., *Kinetics of Wood Sacharification: hydrolysis of Cellulose and Decomposition of Sugars in Dilute Acid at High Temperature.* Industrial & Engineering Chemistry Research, 1945. **37**(1): p. 43-52.

3. Jacobsen, S.E. and C.E. Wyman, *Cellulose and Hemicellulose Hydrolysis Models for Application to Current and Novel Pretreatment Process.* Applied Biochemistry and Biotechnology, 2000. **84-86**: p. 81-96.

4. Yat, S.C., A. Berger, and D.R. Shonnard, (2008) *Kinetic Characterization for Dilute Sulfuric Acid Hydrolysis of Timber Varieties and Switchgrass.* 99(9), pp 3855-3863.

5. Jensen, J.R., Morinelly, J., Aglan, A., Mix, A., Shonnard, D.R., Kinetic Characterization of Biomass Dilute Sulfuric Acid Hydrolysis: Mixtures of Hardwoods, Softwood, and Switchgrass, *AIChE Journal*, online April, 2008, http://www3.interscience.wiley.com/journal/109931315/issue

6. Chen, R., Y.Y. Lee, and R. Torget, *Kinetic and Modeling Investigation on Two-Stage Reverse-flow Reactor as Applied to Dilute-Acid Pretreatment of Agricultural Residues.* Applied Biochemistry and Biotechnology, 1996. **57/58**: p. 133-146.

See more of this Session: Chemical and Biological Processes for Woody Biomass Conversion to Fuels and Chemicals - I

See more of this Group/Topical: Forest and Plant Bioproducts Division

Liquid Products of Liquid-Phase Pyrolysis

Nikolaus Schwaiger, Department of Chemical Engineering and Environmental Technology, Graz University of Technology, Graz, Austria

Verena Mertlitz, Department of Chemical Engineering and Environmental Technology, Graz University of Technology, Graz, Austria

Peter Pucher, BDI BioDiesel International AG, Grambach, Austria

Edgar Ahn, BDI BioDiesel International AG, Grambach, Austria

Matthäus Siebenhofer, Department of Chemical Engineering and Environmental Technology, Graz University of Technology, Graz, Austria

Ligno-cellulosic feed is expected to contribute significantly to production of liquified combustibles in future because of the quantity and the variety of feed material. Therefore the pyrolytic degradation properties of several feed material such as wood-chips, straw and cellulose fibers were investigated. The process was carried out in a semi-batch reaction vessel under isothermal conditions at a temperature of T = 623 K and at ambient pressure. For optimum heat transfer the pyrolysis process was carried out in liquid carrier phase. Gaseous products as well as condensibles were removed on top of the reaction vessel. Condensation temperature was T = 20 °C. After recording of the flux and the composition the gaseous phase was incinerated in a flare. The condensed products were analysed by GC-MS and NMR spectroscopy as well as elemental analysis.

The product yield was determined. Figure 1 shows the summary of water and liquid CHO-products formed during pyrolytic degradation of several feed material.

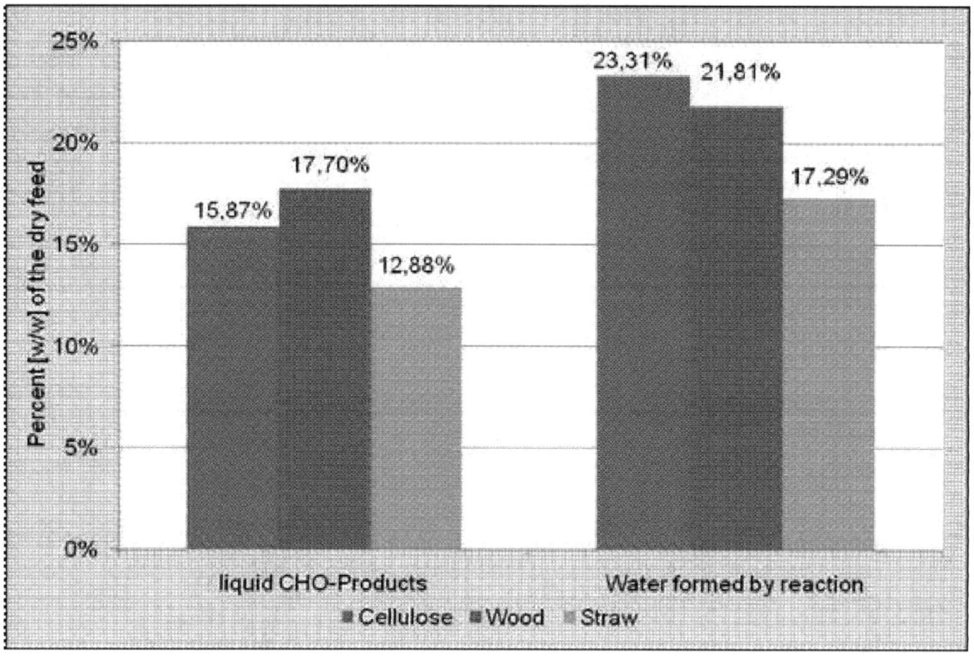

Figure 1: Formation of liquid products by liquid phase pyrolysis

Figure 2 shows the yield of fingerprint products generated by liquid-phase pyrolysis of different feed material.

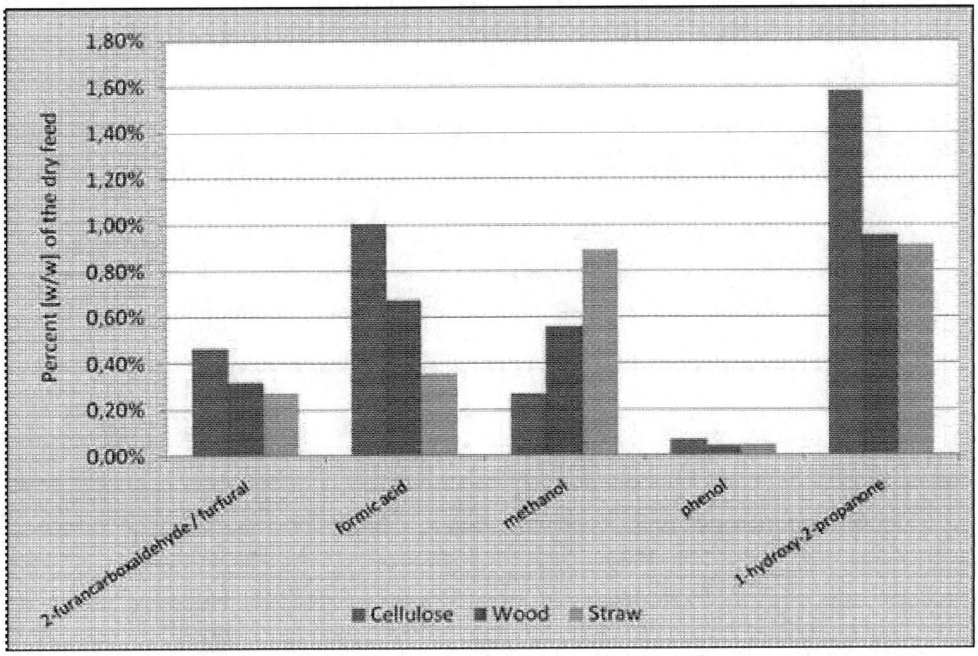

Figure 2: Formation of condensible fingerprint products of liquid phase pyrolysis

The amount of condensibles does not differ dramatically for different feed material. Degradation of cellulose is seemingly controlled by cleavage plus dehydration while ligno-cellulosic feed leads to significant formation of methoxy-substituted phenols.

See more of this Session: Chemical and Biological Processes for Woody Biomass Conversion to Fuels and Chemicals - I

See more of this Group/Topical: Forest and Plant Bioproducts Division

Fast Pyrolysis of Pine Wood: The Influence of Process Parameters on the Quality and the Quantity of the Pyrolysis Oil

Sedat H. Beis[1], Saikrishna Mukkamala[1], Nick Hill[1], Heini Lehtonen[1], Ta-Hsuan Ong[3], Adriaan Van Heiningen[1], Brian G. Frederick[2,4], M. Clayton Wheeler[1], Elizabeth Stemmler[3] and William J. DeSisto[1,4]

[1]Chemical and Biological Engineering, University of Maine, Orono, ME
[2]Department of Chemistry, University of Maine, Orono, ME
[3]Department of Chemistry, Bowdoin College, Brunswick, ME
[4]Laboratory for Surface Science and Technology, University of Maine, Orono, ME

ABSTRACT

In this research fast pyrolysis of pine saw dust was studied to establish the effect of the process parameters such as pyrolysis temperature, feed rate and residence time on conversion and the quality of the products on bench scale. Fast pyrolysis experiments carried out in a bubbling fluid bed reactor heated externally.100g of PSD was pyrolyzed at 400C, 500C, and 600C. The highest conversion yields of liquid, solid and gas products (at respective pyrolysis temperatures) were 56% (500°C), 20%(400°C) and 58% (600°C). General characteristics of bio-oil such as viscosity, Karl Fischer water content, ash, C, H, O, N and metals were determined.13C NMR spectroscopy was used for the determination of the functional groups present in the oil. Separation and quantification of the gaseous products such as H_2, CO, CO_2 and C_1-C_6 range hydrocarbons were performed by gas chromatography. Char was analyzed for metals and ash content.

INTRODUCTION

The thermal conversion of biomass into fuels and chemicals has been spurred by the interest in sustainable, environmentally friendly energy production. Among the many routes to the thermal conversion of biomass, fast pyrolysis provides a means of producing bio-oil, an intermediate envisioned for the production of fuels and commodity chemicals. In Maine, availability of woody biomass waste generated from the logging and lumber industry that has potential as a sustainable biomass feedstock.

In this research, fast pyrolysis of pine sawdust (PSD) has been performed to evaluate the effects of processing parameters such temperature, feed rate, residence time on the quantity and quality of the pyrolysis, products namely bio-oil, char and gas in a bubbling fluid bed reactor pyrolysis unit on bench scale.

MATERIAL AND METHODS

A batch of pine sawdust was subjected to ash, moisture, metals and particle size distribution analysis prior to pyrolysis experiments were initiated. Elemental composition (C, H, N, S, O) of the PSD was also determined. The pyrolysis unit consisted of a custom designed continuous auger type variable speed biomass feeder, a stainless steel fluid bed reactor (D: 1.75" x L: 12"), a char separator, a double pipe condenser and a custom designed electro static precipitator (ESP). Silica sand was used as the heating medium. The reactor was heated in a three zone furnace operated with PID controllers and the hot product line was kept at 350-400C by using flexible band heaters controlled with a variac. Temperatures in the reactor and in hot product lines were monitored with K type thermocouples before and during the runs. The condenser was cooled by a chiller set at 4°C. In the experiments, 100 g of PSD was pyrolyzed at 400°C, 500°C, and 600°C. The runs were made in duplicate. The mass of liquid and char was determined by weighing the unit of interest before and after the run. The mass of gas was determined by subtraction. Liquid products were collected from the condenser and the ESP and kept separate for component distribution, organics content and KF water content evaluation. The bio-oil was characterized for water content, heating value, viscosity and chemical composition. Gaseous products were analyzed with a SRI MG#3 gas chromatograph (GC). The GC was equipped with molecular sieve, Hayesep-D and MXT-1 columns for separation. Detection was achieved with a flame ionization detector/methanizer (FID) and a thermal conductivity detector (TCD) for analysis of H_2, CO, CO_2, C_1-C_6 paraffins and olefins. Solid product was characterized for elemental and ash content.

RESULTS AND DISCUSSION

According to data obtained for fast pyrolysis of PSD at 400°C, 500°C and 600°C gave the product yields of 51%, 56% and 35% for liquids and 20%, 13% and 7% for char and 29%, 32% and 58% for gas respectively. ^{13}C NMR data revealed that distribution of carbon present in bio-oils were in the range of 7.1-4.5 % for carbonyl (215-163ppm), 30.1-50.5% for aromatics (163-110ppm), 3.9-6.3% for carbohydrate (110-84ppm), 23.7-35.9% for methoxy/hydroxy (84-54ppm), 17.3-22.8% for alkyl hydrocarbons (54-1). Bio-oils obtained from runs having 40, 60 and 80 mL fixed bed sizes had 6.0-6.8% carbonyl (215-163ppm), 34.8-35.9% aromatics (163-110ppm), 5.1-5.8% carbohydrate (110-84ppm), 29.5-33.6% methoxy/hydroxy (84-54ppm), 18.0-21.0% alkyl hydrocarbons (54-1ppm).

The data obtained from pyrolysis runs made at 400°C, 500°C and 600°C showed that temperature had the dominant effect with regard to conversion and component profile of the products over feed rate, nitrogen flow rate and bed volume within the ranges studied. The highest conversions obtained in favor of liquids at 500°C (56%), char at 400°C (20%) and gas at 600°C (58%) yielded. It was found that aromaticity of the bio-oils increased as the temperature increased. The content of aromatic C showed a step increase in the bio-oil obtained at 600°C. The gas composition varied considerably depending on the temperature.The gases were present in the range of 45-66% for CO, 19-51% for CO_2, 4-10% for CH_4 and 0-5% for C_2-C_4. No H_2 was detected for the temperature range in this study.

2D HSQC NMR Analysis and Application to Hot-Water Wood Extract Hydrolysate Characterization

Alan M. Shupe, Paper and Bioprocess Engineering, SUNY College of Environmental Science and Forestry, Syracuse, NY

David Kiemle, Chemistry, SUNY College of Environmental Science and Forestry, Syracuse, NY

Shijie Liu, Paper and Bioprocess Engineering, SUNY College of Environmental Science and Forestry, Syracuse, NY

Hot-water extraction has been evolved into a necessary biorefinery step that enables the incremental conversion of lignocellulosic materials to chemicals, materials and energy. The extraction liquor and subsequent hydrolysate are a complex mixture of organic and inorganic substances. 5-carbon and 6-carbon sugars and their derivatives make up the bulk of the mixture. Developing a reliable technique to quantify the mixture has become the bottleneck of the hot-water extraction process development. In this study, we present the development of the 2D HSQC NMR characterization of the complex mixture. It can economically produce quantifications for the monomoeric and some of the oligomeric sugars.

See more of this Session: Chemical and Biological Processes for Woody Biomass Conversion to Fuels and Chemicals - I
See more of this Group/Topical: Forest and Plant Bioproducts Division

Hot-Water Extraction of Aspen Woodchips Prior to Pulping

Houfang Lu, Chemical Engineering, Sichuan University, Chengdu, China

Thomas E. Amidon, SUNY College of Environmental Science and Forestry, Syracuse, NY

Shijie Liu, Paper and Bioprocess Engineering, SUNY College of Environmental Science and Forestry, Syracuse, NY

This paper discusses the kinetic of hot-water extraction of Aspen woodchips. The Aspen woodchips are obtained from Alberta Pacific Forestry Industries, Inc., Boyle, Alberta. The extraction was carried out in hot-water at various temperatures and wood to water ratios. It has been observed that the level of extraction increases with increasing temperature and time. The wood to liquor ratio has little effect on the level of extraction for the range of parameters studied. Further studies were carried out for the pulping of the hot-water extracted woodchips.

See more of this Session: Chemical and Biological Processes for Woody Biomass Conversion to Fuels and Chemicals - I
See more of this Group/Topical: Forest and Plant Bioproducts Division

Wood-Based Biorefinery: An Update On the ESF Technology

Thomas E. Amidon, SUNY College of Environmental Science and Forestry, Syracuse, NY

Shijie Liu, Paper and Bioprocess Engineering, SUNY College of Environmental Science and Forestry, Syracuse, NY

The SUNY ESF biorefinery, a hardwood focused water-based technology, includes five main components or essential elements, namely: 1) hot-water extraction; 2) hydrolysis of hot-water extracts; 3) fractionation of hydrolyzates or wood extracts; 4) conversion of five- and six-carbon sugars to ethanol, butanol, bioplastics, or other products; 5a) pulping of hot-water extracted woodchips followed by bleaching and papermaking; 5b) burning of the hot-water extracted residual chips and organics for CHP use; 5c) pelletizing the hot-water extracted residual chips together with the lignin separated from wood extracts; or 5d) production of fiber board from the hot-water extracted woody biomass. Hot-water extraction in the absence of mineral acids and bases is preferred due to its ability to maintain high product values. Extraction and hydrolysis produce xylo-oligomers, acetic acid, sugars, and other trace compounds. The production of acetic acid makes the process autocatalytic as acid catalyzes extraction and hydrolysis. Membrane fractionation can be applied to purify and concentrate sugars and acetic acid. There is an enhancement in manufacturing processes and several products based the extracted chips, most notably in pulping and bleaching processes. Therefore water based technology is preferred for a wood based biorefinery.

See more of this Session: Chemical and Biological Processes for Woody Biomass Conversion to Fuels and Chemicals - II
See more of this Group/Topical: Forest and Plant Bioproducts Division

Process Engineering for Lignocellulosic Ethanol Facilities: a Combined Model- and Experiment-Based Approach and Comparison with Operational Results From a Demonstration-Scale Plant

Keith Flanegan, KL Energy Corporation, Rapid City, SD

Dave Litzen, KL Energy Corporation, Rapid City, SD

Dennis Harstad, KL Energy Corporation, Rapid City, SD

Jim Schultze, KL Energy Corporation, Rapid City, SD

The Energy Independence and Security Act (EISA) of 2007 mandates that 16 billion gallons per year of cellulosic ethanol shall be produced by 2022. This will require the construction of a large number of facilities using a wide variety of cellulosic ethanol feedstocks. This presents a large technical and economic challenge that can be addressed, in part, by intesified process engineering.

Process design for lignocellulosic ethanol plants is more complicated than for corn-ethanol facilities because the feedstocks have much more compositional variability than common yellow dent #2 corn. Within similar feedstock groups such as coniferous softwoods there is significant variation in the amount of cellulose, hemicellulose, lignin, extractives, acetyl, and ash components, as well as variation in optimal pretreatment severity. Wide variation has been observed even within the same species, depending on the harvest location, age of the plant, and part of the plant that has been harvested. As a result, each new lignocellulosic ethanol facility will typically have a unique feedstock composition and will need significant process design modifications to operate optimally.

Successful, rapid facility design and rapid retrofitting requires a combination of specific process experiments with a robust process model. This paper will describe a methodology to quickly and accurately tailor an entire-facility process design for a unique feedstock or blend of feedstocks. The method has also been applied to convert an existing facility from one type of feedstock to another.

Results from the methodology will be compared with actual operations data from a working, 1.5 million-gallon-per-year lignocellulosic ethanol facility in Upton, WY using a variety of softwood feedstocks.

See more of this Session: Chemical and Biological Processes for Woody Biomass Conversion to Fuels and Chemicals - II
See more of this Group/Topical: Forest and Plant Bioproducts Division

Chemical Characterization of Catalytic Hydrodeoxygenated Bio-Oil and Their Distilled Fractions

El barbary M. Hassan, Forest Products, Mississippi State University, Mississippi State, MS

Sanjeev K. Gajjela, Forest Products, Mississippi State University, Mississippi State, MS

Fei Yu, Forest Products, Mississippi State University, Mississippi State, MS

Phillip Steele, Forest products, Mississippi State University, Mississippi State, MS

Fast pyrolysis is a potentially important route for biomass conversion to fuels. Researchers have shown that the problematic raw bio-oil properties of high acidity, high water content, lower heating value and variable viscosity are resolved by application of hydrodeoxydation (HDO). In this study, we pyrolyzed pine wood in an auger reactor at 450 C and the produced bio-oil was upgraded by HDO single-stage catalysis. The hydrocarbons produced in the HDO bio-oil were in the range between gasolin and Jet-fuels. We distilled the HDO bio-oil to determine the ability of producing relatively clean hydrocarbon fractions. The HDO bio-oil was distilled by spinning band distillation column into fractional boiling point ranges. The upgraded bio-oil and distilled fractions were characterized by GC/MS, FT-IR, NMR and elemental analysis. The distilled fractions were also compared with transport fuels by simulated distillation. Physical properties, such as higher heating value, viscosity, pour and flash points were also determined.

See more of this Session: Chemical and Biological Processes for Woody Biomass Conversion to Fuels and Chemicals - II
See more of this Group/Topical: Forest and Plant Bioproducts Division

Biomass Conversion Via Direct Chemical Looping Technology – Process Simulations

Liang Zeng[1], Fanxing Li[1], Rae Hyung Kim[1], Deepak Sridhar[1], Fei Wang[1], Andrew Tong[1], Zhenchao Sun[1], Kobayashi Nobusuke[2] and L. -S. Fan[1]

1. William G. Lowrie Department of Chemical and Biomolecular Engineering
The Ohio State University
2. Department of Chemical Engineering
Nagoya University

Abstract

Biomass is quite a promising renewable energy source due to its clean nature and abundant reserves. Conventional processes for biomass conversion face several limitations, which may hold back their further development. Chemical looping technology is an effective thermochemical strategy for hydrogen and electricity coproduction from biomass. In this paper, a process evaluation is conducted for biomass conversion through iron-based direct chemical looping technology. The biomass direct chemical looping concept and its overall energy management are elaborated. A multistage model is set up in Aspen Plus® to study thermodynamic equilibrium in the moving bed reducer. The model can help to predict the performance of moving bed reactor and optimize the operating conditions. An overall process simulation is then developed with predicted reactor performance and reasonable assumptions in Aspen Plus®. Simulation results show that biomass direct chemical looping (BDCL) is quite efficient in biomass conversion to hydrogen and/or electricity with simple control. Several scenarios are studied for process optimization. Finally, some technical considerations are addressed for further discussion on biomass direct chemical looping technology.

1. Introduction

Woody biomass is a clean and renewable energy source. This renewable resource can be converted into electricity, H_2, chemicals, and liquid fuels. Compared to fossil fuel, biomass conversion is less carbon and pollutant intensive. Moreover, it is widely distributed and abundantly available. For instance, the annual biomass potential from forest and agricultural resources alone is over 1.3 billion dry tonnes in the United States, which is sufficient to displace nearly 40% of the petroleum consumed in the nation [1]. Therefore, biomass has the potential to be a favorable energy source to, at least partially, address the increasing environmental and energy sustainability concerns.

The chemical looping technology [2] convert carbonaceous fuels using metal-oxide based oxygen carriers circulating between two main reactors, a reducer and an oxidizer. Within the redox loop, oxygen transfers from the metal oxides to fuel in the reducer and from steam or air to reduced metal oxides in the oxidizer. This strategy avoids direct contact between fuel and oxidants. Hence, a separate CO_2 product stream is generated with minimal energy consumption. The direct conversion of solid fuels such as biomass using chemical looping strategy is a relatively new subject. The biomass direct chemical looping (BDCL) process developed at OSU is a promising technology to combine hydrogen production and a readily sequestrable CO_2 stream in situ. As shown in Figure 1, BDCL utilizes the chemical looping strategy to directly convert biomass into hydrogen. Unlike the gasification with oxygen, biomass is indirectly gasified by a patented oxygen carrier composing mainly of iron oxides with some support compounds. A concentrated CO_2 stream can be obtained in this process, obviating the CO_2 separation step. The reduced form of iron oxide can then be reacted with air to liberate heat to produce electricity or reacted with steam to form a relatively pure stream of hydrogen, which can then be used for a variety of purposes.

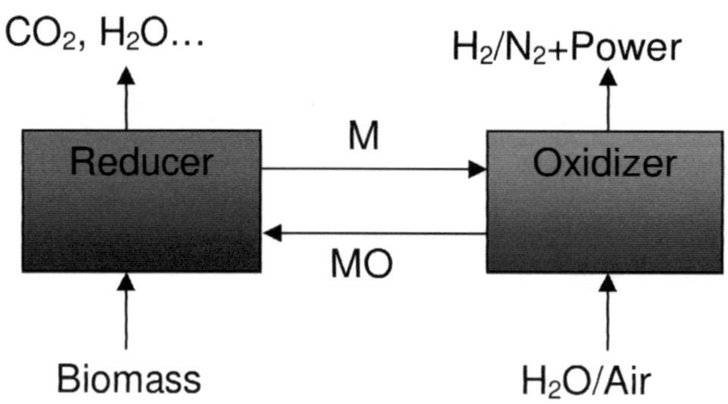

Figure 1 Simplified Schematic of Chemical looping strategy

The present work is to develop a conceptual design for the BDCL process. Each reactor is described in detail, and then integrated together through energy management. Novel thermodynamic models for the moving bed reactor in the BDCL system are set up and used for analysis in the BDCL system. The calculated performance elicits several discussion topics. Process simulations using Aspen Plus® are developed for further illustration based on thermodynamic analysis.

2. Conceptual Design

This section illustrates one schematic configuration for the BDCL process. There are three major units in the BDCL, i.e. Reducer (1), Oxidizer (2), and Combustor (3). A simplified diagram for the BDCL process is shown in Figure 2.

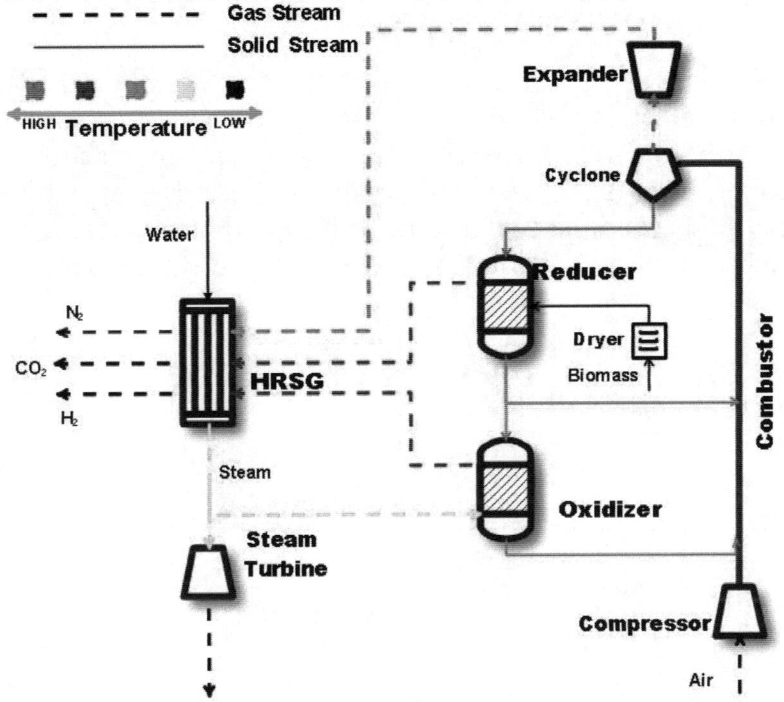

Figure 2 A simplified flow diagram for Biomass Direct Chemical Looping process

Reducer

The reducer is a moving bed reactor where the oxygen carrier particles, in Fe_2O_3 form, react with biomass feed (considered as carbon in this section) to produce reduced Fe/FeO solids and a gaseous CO_2/H_2O stream. Here, the Fe_2O_3 particles are fed from the top while the dried biomass is injected into the middle section of the reactor. Additionally, a small amount of gasifying agents such as steam, CO_2, and/or H_2 can be used to enhance the reaction kinetics by gasifying the solid fuel. The reducer operates at 850 °C~950 °C and atmospheric pressure. The reactions can be represented as:

$3Fe_2O_3+CO—2Fe_3O_4+CO_2$ (1)
$Fe_3O_4+CO—3FeO+CO_2$ (2)
$FeO+C—Fe+CO$ (3)

Oxidizer

A portion of the reduced oxygen carrier particles from the reducer outlet are transferred to the oxidizer for hydrogen production. The remaining reduced oxygen carrier particles are directly combusted with air in the combustor for heat generation. The oxidizer is a countercurrent moving bed reactor that produces hydrogen via the steam-iron reaction at 700 °C~900 °C and 30 atm. Fe and FeO are oxidized to Fe_3O_4 while H_2O is reduced to H_2 as shown in reactions (4) and (5). These reactions are mildly exothermic.

$Fe+H_2O—FeO+H_2$ (4)

$3FeO+H_2O—Fe_3O_4+H_2$ (5)

Combustor

The combustor is an entrained flow reactor through which the oxygen carrier, from either the reducer or the oxidizer, is transferred back to the reducer by compressed air. During this step, the oxygen carrier is reoxidized to Fe_2O_3 through reaction (6-8) where a significant amount of heat is released. The solids can reach a temperature of more than 1150 °C. The hot solids from the combustor can fully compensate the heat needed in the reducer. The high quality heat within the flue gas can drive expanders for electricity generation.

$4Fe_3O_4+O_2—6Fe_2O_3$ (6)

$4FeO+O_2—2Fe_2O_3$ (7)

$4Fe+3O_2—2Fe_2O_3$ (8)

When the looping system is considered as a whole, the overall reaction can be expressed as:

$$xC+yH_2O+(x-y/2)O_2—xCO_2+yH_2+\Delta H$$

Thermodynamically, the net heat of the system can be adjusted to zero by changing the ratio between x and y. However, the system is often adjusted to be exothermic so that steam and power can be co-produced at the expense of the hydrogen yield. Figure 3 summarizes a schematic for the material flow and energy flow circulated in this configuration.

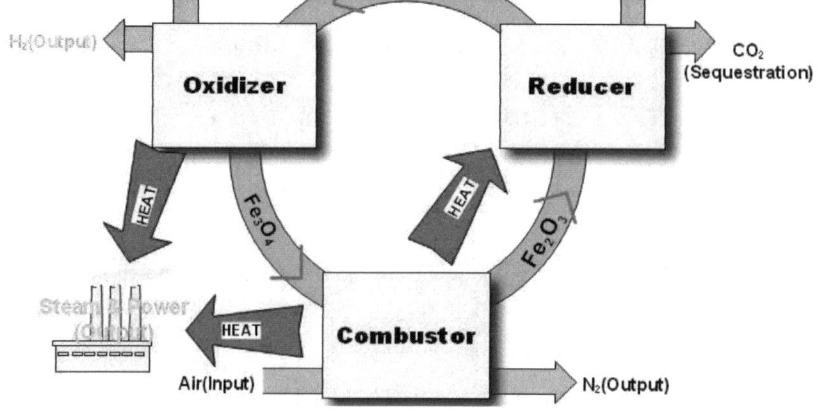

Figure 3 material flow and energy flow in BDCL process

3. BDCL Reducer Modeling

Thermodynamic restrictions are the first and foremost thing to check for a process development, especially reactor design in the BDCL system. As described in previous section, moving bed reducer and oxidizer are adopted in BDCL system, which is unusual compared with previous studies. The gas-solid countercurrent flow pattern in the moving bed reactor has the potential to enhance reaction thermodynamics, as it maximizes reaction driving force along the reactor. It is more advanced than a fluidized bed design where products and reactants are mixed together, resulting in weakened driving force and poor performance. In order to predict the performance of the moving bed reactor, thermodynamic analysis on the reducer is performed based on ASPEN Plus® software.

Model Setup

For the simulation of countercurrent moving bed, one effective way is to configure a set of multiple interconnected RGibbs blocks as shown in Figure 4. One RYield block first decomposes biomass into its constituent elements based on its ultimate analysis. In Rgibbs blocks, thermodynamic equilibrium among all possible products is calculated based on the principle of minimization in Gibbs free energy. Biomass is injected into the middle stage of the model, while iron oxide is introduced from the top stage. A stream of promoter such as CO_2 or steam can also be introduced from the bottom stage. In the multistage model, gas and solid reactants will reach thermodynamic equilibrium in each stage under the designated conditions. All the gases will flow upwards while the solids will move downwards throughout these stages. The following simulation results have been compared with both ongoing experimental studies and thermodynamic analysis in literatures, showing relatively good accuracy.

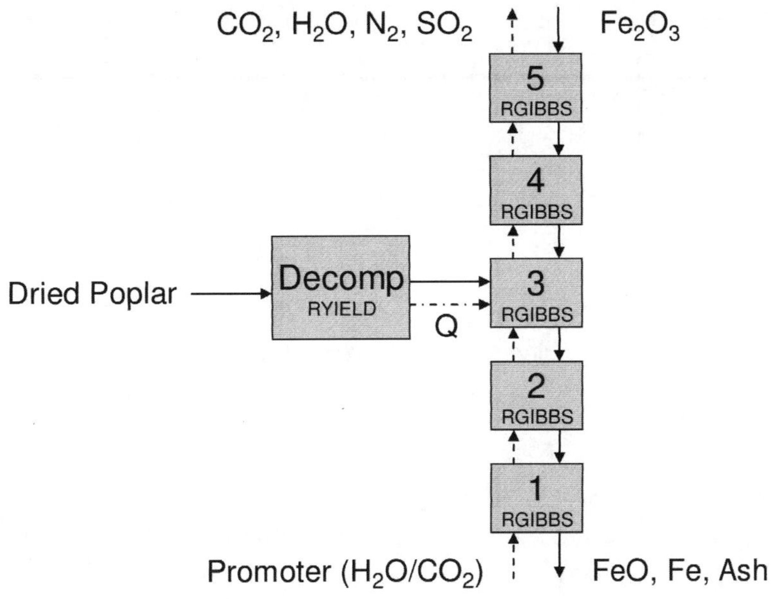

Figure 4 Model setup for moving bed reducer

Results and discussions on reducer modeling

For quantification, a set of parameters are defined firstly:
M = mass percentage of moisture in dried biomass
R = molar ratio between fresh Fe_2O_3 in oxygen carrier and carbon in biomass in the feedstock
Oxygen carrier conversion = 1- [O] in reduced oxygen carrier/ [O] in fresh oxygen carrier

Effect of moisture

An ideal reducer should be configured so that biomass is fully oxidized into CO_2 and H_2O, and then there should be a minimum R. At 900 °C and 1 atm, the developed model is carried out in this situation to compare the performances with different feed-in moisture. The results are illustrated in Figure 5, where E curve represents the ratio of heat required for constant temperature operation relative to 5% moisture operation. It shows that, for non-drying biomass input (M = 50%), at 900 °C and 1 atm, the maximized oxygen carrier conversion in a moving bed reducer is 26.8% with R min = 2.60. As the moisture is dried out, both R and E drops while the C (oxygen carrier conversion) increases. It means that the lower moisture content in feedstock, the lower recycle rate and energy consumption. Similar effect is found for the injection of steam or CO_2 as a promoter. Therefore, it is always good to dry biomass first in BDCL process.

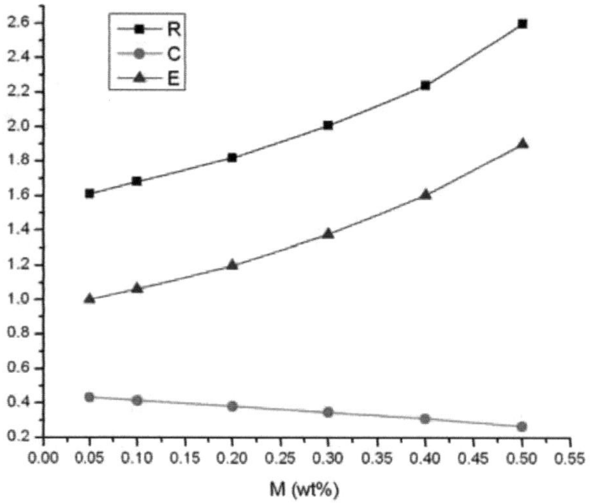

Figure 5 Effect of moisture on moving bed reducer

Products distribution

With M=5%, R=1.61 at 900 °C and 1 atm, the resulting gas and solid stream is summarized in Table 1. The model predicts that biomass is fully converted to CO_2 and H_2O while iron oxide is reduced by 43.4%. All the sulfur will exit from the reducer with the gaseous stream in the form of SO_2. No sulfur will accumulate in oxygen carrier due to its low concentration in biomass. This indicates that all the sulfur can be recovered from the gas stream out of reducer using state-of-art technology.

Table 1 Yields from moving bed reducer with M=5%, at 900 °C, 1 atm.

Gas Stream, mol%		Solid Stream, mol% (Ash Free)	
CO_2	0.562306	$FE_{0.947}O$	0.812341
H_2O	0.436501	FE	0.187659
N_2	0.000806	$Fe_{0.877}S$	0
SO_2	0.000373		

Effect of Temperature

Sensitivity analysis can be carried out using the multistage equilibrium model for further study. The carbon distributions in reaction products with constant pressure at 1 atm and varying temperature are shown in Figure 6. From thermodynamic viewpoint, a higher reaction temperature favors the endothermic biomass-Fe_2O_3 reaction, which is also predicted by the model. In Figure 6, using a Fe_2O_3/biomass ratio identical to the case shown before (M = 5%, R = 1.61), carbon can not be fully converted at temperatures below 900 °C unless increasing R. Therefore, the reducer should work at high temperature to maximize both the fuel and oxygen carrier conversion.

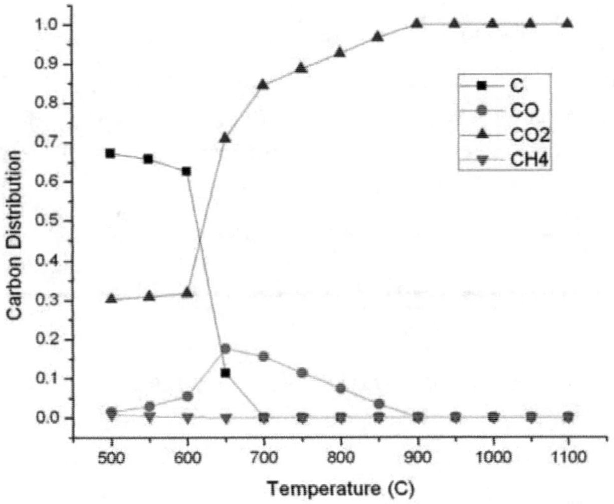

Figure 6 Effect of temperature on biomass conversions at 1 atm

Effect of Pressure

The reactions involved in the reducer generates gaseous products, tending to increase the overall volume of the system. Therefore, a higher operating pressure inhibits the desired reaction scheme. As shown in Figure 7 (M = 5%, R = 1.61, 900 °C), low pressures will favor biomass conversion, whereas high pressures will block the carbon gasification. It is determined that an operating pressure range between 1 atm and 30 atm will be suitable for the BDCL reducer operations. However, if considering the downstream CO_2 compression instead of direct ventilation, it is appropriate to operate the reducer at higher pressure.

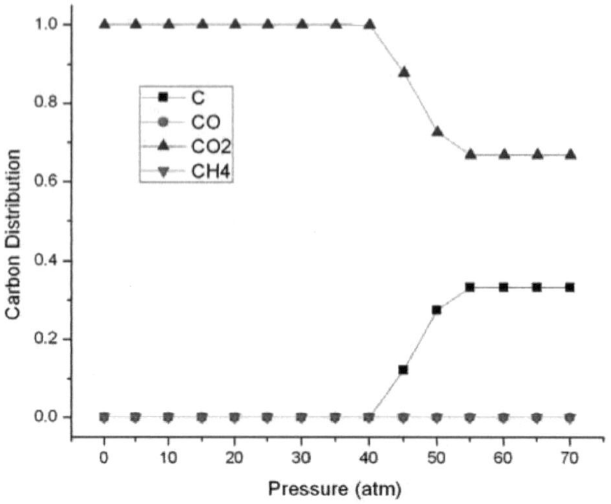

Figure 7 Effect of pressure on biomass conversion at 900 °C

To sum up, through the developed multistage model, the operating conditions in a moving bed reducer can be forecasted and optimized. Similar model can be used to analyze the performance in the oxidizer with countercurrent gas solid contacting pattern. For the entrained bed type combustor, one stage RGibbs model can be used for estimation. Using those reactor models, thermodynamic limitations for the BDCL system can be clarified for further process analysis and optimization.

4. Process Simulation on BDCL process

Grounded on thermodynamic analysis, a conceptual process design is constructed in this section for hydrogen production via BDCL process. As shown in Figure 8, BDCL process will mainly embrace biomass preparation, DCL system, heat recovery and steam generation (HRSG) system, gas cleanup units and power system.

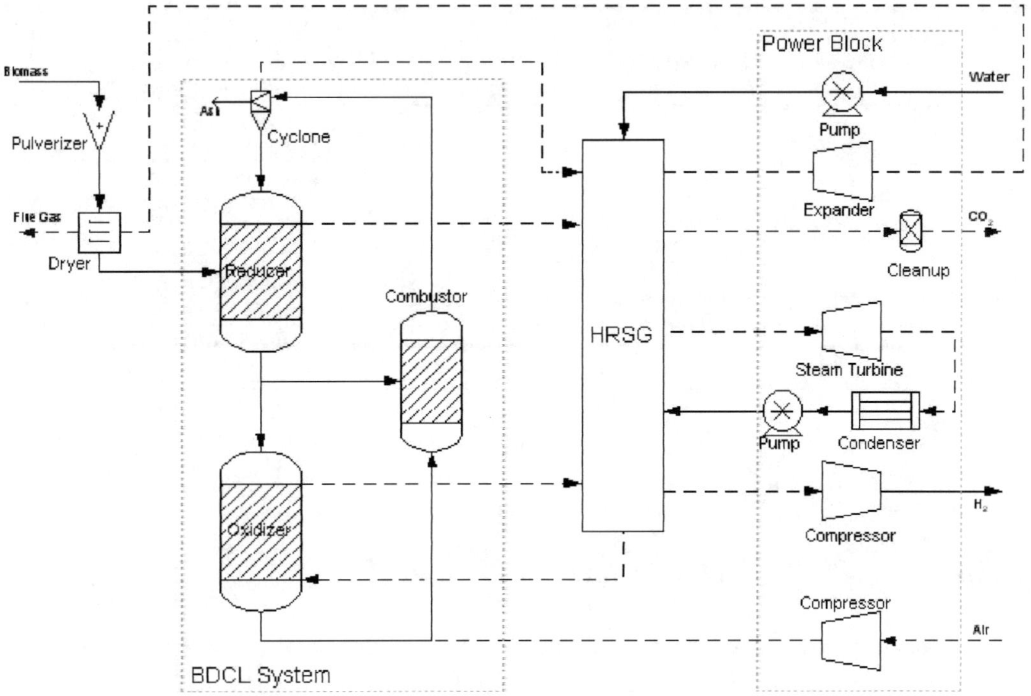

Figure 8 Process Flow Diagram for BDCL Process for Hydrogen Production

Biomass Preparation

Biomass preparation section involves mechanical pulverization and flue gas drying. As mentioned before, the moisture content will not only increase the energy waste by water vaporization, but also decrease the reducer performance. Therefore, it must be dried out as much as possible before entering reducer. In this case, the poplar is first crushed and pulverized to fine power (~5 mm) using KDS mills, in the meantime, the water content can be removed to about 20% by mechanical action. For current capacity of 100 MW thermal inputs, the pulverization shares about 2.86 MW of electricity from the gross power output. Then the pulverized poplar is dried to around 5% moistures by a stream of warm flue gas originated from the combustor.

BDCL System

The dried poplar is introduced into the BDCL system from the reducer, performed at 900 °C, 1 atm. 192.64 tonne/hr of Fe_2O_3, together with 98.2 tonne/hr SiC, is introduced from the top, and then reduced to form of Fe and $Fe_{0.947}O$ (wuestite) as converting all the fuels. All the sulfur exits with the gas phase as SO_2. 65.7% of the reduced iron is directed to the reducer for hydrogen production, and the other 34.3% goes to combustor directly for heat balance. The oxidizer, whereafter, carries out the iron-steam reaction using the reduced solids at 850 °C, 30 atm. The steam for hydrogen production flows at 25.52 tonne/hr, and has a conversion of 54.7% predicted by multistage oxidizer model. The combustor regenerates Fe_2O_3 at 1324 °C, 30 atm. After the looping, poplar is fully converted and automatically separated to CO_2 stream (consisting of mainly 44% CO_2 and 56% steam), H_2 stream (about 55% H_2 and 45% steam). More simulation results are listed in Table 2.

Table 2 Simulation Results in BDCL system

Reactor	Feedstock Condition	Operating Pressure,	Operating Temperature	Mole Fraction in Products (excluding ash)	Conversion
Reducer	Dried Poplar	1 atm	900 °C	Gas: 0.4365 H_2O, 0.5623 CO_2, 806ppm N_2, 373ppm SO_2	Biomass: 100%
	25 °C, 1 atm			Solid: 0.097 Fe, 0.410 $Fe_{0.947}O$, 0.493 SiC	Fe_2O_3: 43.4%
Oxidizer	Steam	30 atm	850 °C	Gas: 0.453 H_2O, 0.547 H_2	Steam: 54.7%
	240 °C, 32 atm			Solid: 0.247 Fe_3O_4, 0.753 SiC	To Fe_3O_4: 100%
Combustor	Compressed Air	16 atm	1324 °C	Gas: 0.01 O_2, 0.989 N_2, 509ppm NO	O_2: 96.2%
	126 °C, 18 atm			Solid: 0.33 Fe_2O_3, 0.67 SiC	To Fe_2O_3: 100%

Heat Recovery and Steam Generation

In BDCL process, most of the reactions happen at high temperature, resulting in a number of hot streams. The HRSG unit is used to recover the available heat from these gaseous streams. About 15.99 tonnes/hr of high temperature high pressure steam, will be raised in HRSG to drive steam turbines, while another low pressure steam, 25.52 tonne/hr at 240 °C and 32 atm, will be the feedstock for the oxidizer. All the gas will be cooled down to 120 °C before leaving HRSG unit.

Power Generation

The major energy consumptions in BDCL process comes from gas compressors and biomass pulverizer. An expander is used to generate power to compensate the parasitic energy using the hot flue gas from combustor. The superheated high pressure steam from HRSG can be also used for electricity generation through steam turbines. The power balance is summarized in Table 3.

Table 3 Power balance in BDCL process

Unit Operations	Input				Output				Net power
	Air compressor	H2 compressor	Pulverizer	other	HP steam turbine	IP steam turbine	LP steam turbine	Expander	
Power, MW	5.52	0.99	2.86	2.21	-1.74	-3.09	-1.99	-10.31	-5.55

Process Efficiency

We define the efficiency (HHV) for the process as follows,

η_e = Net Power (MW)/ Higher Heating Value in Biomass Feedstock (MW)

ηH_2 = Higher Heating Value in H_2 Procect (MW)/ Higher Heating Value in Biomass Feedstock (MW)

$\eta tot = \eta e + \eta H_2$

Based on this conservative simulation, about 1.56 tonne/hr compressed hydrogen is produced from BDCL process with $\eta H_2 = 61.60\%$. After balancing the parasitic energy, there is also about 5.55 MW of net electricity available for grid. Therefore, the BDCL process can coproduce hydrogen and electricity with an efficiency as high as $\eta tot = 67.15\%$.

5. Conclusions

To conclude, BDCL process can coproduce hydrogen and electricity from biomass with considerably increased efficiency and no harm to environment. The drastic improvement roots in its advanced concept and unique energy management. A multistage model has been developed for thermodynamic analysis on moving bed reducer. Based on predicted data from reactor modeling, process simulation is conducted using Aspen Plus, proving that BDCL process represents a clean, high effective and low cost energy utilization structure in biomass conversion.

Reference:

1. Perlack R.D., Wright L.L., Turhollow A.F., Graham R.L., Stokes B.J., and Erbach D.C., Biomass as feedstock for a bioenergy and bioproducts industry: The technical feasability of a billion-ton annual supply. U.S. DOE & USDA, 2005: 16-18.
2. Liang-Shih Fan, F.L., Shwetha Ramkumar, *Utilization of chemical looping strategy in biomass gasification processes.* China Particuology, 2008.(2008, 6(3)): p. 12.
3. AspenTech, *Getting Started Modeling Processes with Solids*. 2006.

Ethanol Fermentation From Hot-Water Wood Extract Hydrolysate with A Modified Pichia Stipitis Strain HWR

Alan M. Shupe, Paper and Bioprocess Engineering, SUNY College of Environmental Science and Forestry, Syracuse, NY

Shijie Liu, Paper and Bioprocess Engineering, SUNY College of Environmental Science and Forestry, Syracuse, NY

Hot-water extraction can be viewed as an efficient pre-treatment technology for woody biomass bioconversion. Promoting the extraction to a process step rather than a pre-treatment step significantly improves the outlook of the overall process. The acid hydrolysate of the wood extracts contains mainly 5-carbon sugars. Our earlier studies has shown that Pichia stipitis NRRL Y-11543 was promising in utilization the mixed sugars in the wood extract hydrolysate. However, NRRL Y-11543 was inhibited in dilute sulfuric acid hydrolysate. Repeated adaptation of Pichia stipitis strain NRRL Y-11543 were carried out to improve the ethanol production efficiency for conversion of hot-water wood extract hydrolysates. Fermentation experiments indicated that fbwhr was resistant to the toxicity of hydrolysate in the fermentation media of concentrated hydrolysate, and xylose was completely utilized by the strain to produce ethanol. HWR was grown in the concentrated hydrolysate without any treatment and yielded ~40 g/L ethanol.

See more of this Session: Chemical and Biological Processes for Woody Biomass Conversion to Fuels and Chemicals - II
See more of this Group/Topical: Forest and Plant Bioproducts Division

Anaerobic Fermentation of Glucose, Xylose and Hemicellulose Present in Pre-Pulping Extracts of Northern Hardwoods to Carboxylic Acids.

Rakhi Reddy Baddam, G. Peter van Walsum

University of Maine, Department of Chemical and Biological Engineering, Orono, Maine 04469. USA

Abstract

Wood is composed primarily of cellulose, hemicellulose and lignin. In the paper industry the cellulose fraction is important as the major resource in paper production, whereas the hemicellulose and the lignin are usually burned for heat recovery. Recently the hemicellulose has received much attention for the production of different bioproducts. Varying amounts of hemicellulose and associated organic acids can be extracted from wood chips when they are subjected to pre-pulping extraction under different conditions of chemical loading and extraction severity. A major interest for use of the extract this is to convert the hemicellulose into valuable chemicals via fermentation.

Mixed-culture microbial ecosystems operated in acidogenic mode are capable degrading all sorts of biomass into mixtures of carboxylic acids. Such cultures were applied to the conversion of green liquor and hot water extracts derived from northern hardwoods. In this study, we determined the production levels of carboxylic acids using mesophilic and thermophilic cultures with green liquor, ammonium bicarbonate and calcium carbonate as the buffering agents. Conditions of combined extraction and fermentation conditions were assessed to determine optimal yields of carboxylic acids for use as platform chemicals for further upgrading. Preliminary results on glucose and xylose demonstrate optimal yield of organic acids with mesophillic culture growing with ammonium bicarbonate buffer.

Introduction

The importance of renewable resources in producing high end products and there by replacing non-renewable resources has been taking in effect in the recent years. This is due to the extensive use of nonrenewable resources to satisfy our present energy needs, leading to the shrinking of these resources over the next few decades. These also lead to a lot of environmental problems. To overcome these problems scientists and researchers are performing extensive research in looking for other resources to replace the existing nonrenewable resources.

There are various renewable resources to replace our present energy demand, including food crops, wind energy, biomass and so on. The use of food sources has a high impact on the economy as the use of these for energy production has the tendency to increase food prices and cause shortages in their availability as food for the community. Another and easily available resource is biomass. The most abundant available biomass such as lignocelluloses are the best resource for the production of fuels and chemicals (Binder et al, 2008). Plant biomass, with abundant lignocellulose, was the source for our energy demands before the discovery of fossil fuels (Huber et al, 2006). We are now moving our efforts towards again making use of the forest resources. One approach to the production of bio-energy and bio-materials is to develop an integrated forest products bio-refinery. It has been proposed that bio-products could be produced by the pulp and paper industry as an additional way to add value to their established industry. (van Heiningen, 2006)

The state of Maine, with 90% of its land covered by forest, is one of the states in the U.S. that produces large quantities of paper and other products from wood. Wood is used as a major resource in Maine to produce various products. While paper is one of the most highly produced products, there are other products derived from wood such as panel boards, building materials and so on.

Wood, also known as lignocellulose, is comprised of cellulose, hemicellulose and lignin. The hemicellulose is the second predominant component next to cellulose in the wood. The wood used in making paper and panel boards is cooked at very high temperatures to remove hemicelluloses and lignin, while retaining the cellulose to provide the strength and quality of the pulp. The wood sugars that can be collected as pre-pulping extracts are predominantly xylo-oligomeric sugars with other sugars and acids(Tunc. M et al).

The major sources for collecting these sugars could be from Kraft pulp mills and possible wood digesters involved in making panel board. The total sugar content in these extracts varies depending on the thermal and chemical gradients used for pre-treating the wood chips prior to pulp processing. The kraft pulp mill can use green liquor to extract a large portion of the hemicellulose and the lignin present in the woodchips while still retaining high pulp yield (Um. et al). Thus the extract generated by using green liquor is called as the Green Liquor Extract. In the panel board making industries these wood chips would be treated with water only, resulting in removal of certain amount of sugars and the lignin. This has one beneficial effect of reducing susceptibility to microbial damage to the wood panels and increases their life time. The extracts thus generated by using water are referred to as Hot water extracts.

Recently there has been much interest shown on these liquid extracts so as to completely use the sugars present in them to produce other useful bioproducts such as volatile organic acids, ethanol, and so on(Balaban 2003, van Heningen 2006). With the use of theses extracts there is a hope to reduce the use of food sources for energy production. The sugars present in aquous wood extracts are dilutely concentrated and contain a high level of acetic acid, which have high value in the market.

Organic acids ranging from C2-C7 are of much interest in the current work. Biodegradable material, especially lignocellulose can be converted in to volatile organic acids by the use of Mixalco process which uses a mixed culture of microbial organisms (Holtzapple et al, 1999). We are applying this method in seeking to convert hemicelluloses in wood extract into organic acids.

Studies are being conducted on $CaCO_3$ and NH_4HCO_3 as potential neutralizing agents to keep the fermentation continuous at a steady pH. The use of methane analogs (eg: Bromoform, iodoform) showed a good results in inhibiting the methane production (Holtzapple et al, 1999). Through this methane inhibition, a higher of carboxylic acid is produced by eliminating the potential hydrogen sink.(Ross. 1998)

To best determine the Mixalco process to be applied directly on the hemicelluloses sugars a standard sugar experiments had first been performed on a batch scale with equal concentration of monomeric glucose and xylose. This medium has no neutrient source so corn steep liquor rich in nutrients is added in the amount of about 3% of the total sugar concentration. The batch experiments are performed anerobically at two different temperatures, one at 37°C and the other at 55°C, to determine the best temperature for a high content of carboxylic acid yields.

II. Materials and Methods

2.1 Substrates

The liquid extracts which is used as the raw material are collected from wood chip digesters located in the process development pilot plant in the Department of Chemical and Biological Engineering at the University of Maine. The digesters are designed to simulate pulping of wood chips in a continuous digester. Extracts used for this study were prepared by carrying out relatively mild condition extractions on hardwoods using either hot water or water mixed with green liquor, which is a highly buffered intermediate pulping stream in the Kraft pulping process.

The green liquor and hot water extraction processes are carried out at varied levels of temperatures, chemical loadings, wood to liquor ratio and periods of reaction time. A typical H factor for the cooking conditions used was on the order of 800.

In green liquor and hot water extracts, the average sugar content in the extracts was noted to be around 8-10 g/L, the acetic acid around 10 g/L. and the average acid insoluble lignin content was measured to be around 5g/L.

2.2 Media and Nutrients

In some experiments, no nutrient supplements were added to the extaction liquors. In other cases, corn steep Liquor (Sigma Aldrich, St Louis, MO) was added at a rate of 3 g/L .

2.3 Buffers.

Calcium Carbonate or Ammonium Bicarbonate is used as a buffering agent , it is being added as a solid form of an amount of 50gms/L of Hot Water extract taken. This helps in maintaining a neutral pH for a better growth of the Microorganisms by neurtralizing the acids produced in the fermentation process. The excess which is not used settles at the bottom and is used as needed. Ammonium bicarbonate dissolves completely upon addition and was added when the fermentation pH droped below 5.8. For a green liquor extract the pH was above neutral and does not require any buffers at the initiation of the experiment.

2.4 Inoculum

Inocula for this experiment were obtained from various sources, including organic sediment from the Great Salt Lake in Utah, secondary clarifier sludge from the waste water treatment facility at a kraft pulp mill and salt water sediment from the coast of Maine. The mixed cultures from these sources were maintained by serial transfer under anaerobic conditions over several months.

2.5 Methanogen Inhibitor

Iodoform(CHI_3) is used as an inhibitor to prevent or diminish the production of methane. In several cases, methane production was low and no inhibitor was required. Excessive use of these chemical may harm the growth of the other microbes present in the mixed culture, and so should be used sparingly. The concentration required for effective methane inhibition is about 10 mg/L. As these compounds are sensitive to light and air it is preferred that they be stored in a tinted glass and capped immediately after use. The iodoform is added as a 20 g/L solution in ethanol. It is preferred to prepare these at the time of use in order to avoid prolonged storage which might affect their stability. When ammonium carbonate is used as a buffer, this will also inhibit methanogenesis.

2.6 Batch Fermentation Process

Experiments are being performed in batch mode to determine optimal productivity from the available sugar content of the extracts. All the batch experiments performed are performed in duplicate/ triplicate to determine reproducibility of results. The batch fermentations are performed in 250 ml serum bottles with a narrow mouth. The mouth is sealed by using a rubber septum stopper and an aluminum crimp which holds the rubber stopper tightly to the mouth of the glass serum bottle. The rubber septa enable collection and measurement the total gas produced during the fermentation by inserting a syringe needle connected to a gas accumulation and volume measurement device.

The bottles are filled with approximately 100ml of the extract in each bottle, which leaves sufficient head space for accumulation of produced gasses. These bottles ate placed in a shaking incubator with continuous shaking at about 150 rpm and constant temperature maintenance. Two different temperatures are investigated for the process: one at 37°C and the other at 55°C.

The gas samples and the liquid samples are taken at regular intervals of time so as to determine the total acid productivity, carboxylic acid concentration, total yield, available sugar concentrations and to check for any presence of methanogens by measuring any methane production during the Process.

2.7 Analytical Methods

2.7.1 Gas Volume Measurement

The total gas volume in the serum vial is measured by connecting it to a specific gas volume measuring device by noting down the total water displacement in the device. This specific instrument is built within the lab using a graduated glass burette, a rubber stopper and few connections. It consists of a 30% Calcium Chloride solution in to a 2 liter beaker, which is used as a gas displacement fluid. The calcium chloride is added so that no carbon dioxide dissolves in to the fluid. A graduated burette which has a capacity to measure 100ml of fluid is taken and the bottom is kept opened. It is arranged such that the bottom tip is completely immense in to the fluid and no leaks are visible when the fluid is taken to the maximum level of the burette. The top of the burette is fitted with a rubber stopper with a small tube through the stopper which is connected to the sampling tube and a vacuum pump, whcich is used to pull the liquid up into the burette before the measurment. The graduations on the tube help in measuring the total volume of the gas produced during the fermentation process.

2.7.2 Gas Analysis

Methane and carbon dioxide concentrations in the produced gas are measured using a SRI instruments multiple gas analyzer, which makes use of three columns including a 60 m MXT-1, a MoleSieve 13X and a HayeSep-D column. Detection is done using TCD, FID with methanizer and FPD detectors.

2.7.3 Sugar Analysis

The samples from the batch cultures are collected at regular intervals of time and these samples are used to analyzed to measure the total sugar content left with in the reactor and also the total volatile solids present within the system. The measurement of the total sugars is performed on a HPLC system by Shimadzu which is equipped with the "Aminex HPX 87H" column supplied by the Bio-Rad Laboratories. The eluent used for this column was 5mM H_2SO_4. The flow rate was maintained at 0.6mL/min through the column. The temperature of the column was set at 60°C. The liquid samples are mixed with fucose as an internal standard and run through the HPLC system. The data is measured using a refractive Index detector set to 40°C. The formic acid in the samples are also measured using the UV- spectrophotometer set at 280nm. The dry solids are measured by drying them in an oven at 40°C for overnight.

2.7.5 Gas Chromatography for measuring acids

The liquid samples collected from the batch cultures are also analyzed on a GC system. Acids in the sample are analysed on a Shimadzu GC-2010 gas chromatograph with a capillary column "Stebilwax-DA" supplied by Restek corporation. It is equipped with Flame Ionisation detector. The samples are injected in to the GC using an auto injector. The column temperature was maintained at 50°C for 1 min and later increased from 50°C to 200°C at a rate of 20°C/min. and kept hole at 200°C for 1min. these samples are mixed with 1.162gm/L of 4-methyl -valeric acid as an internal standard and acidified with 3M Phosporic Acid(H_3PO_4) before injecting in to the GC.

III Results and Discussion

3.1 Batch Fermentation:

Fermentation results are shown in figures 1 – 4. Graphs show consumption of glucose and xylose and accumulation of organic acids and ethanol. All experiments were repeated in quadruplicate. Four conditions were tested, two buffers (ammonium bicarbonate and calcium carbonate) and two temperatures (37 and 55 °C). In all cases, glucose was consumed preferentially to xylose, but all of the mixed cultures did eventually consume the xylose, as well. Organic acids accumulated in all of the fermentations, though lactic acid was usually consumed after initially accumulating. Low pH conditions favored lactic acid production while neutral pH appears to have resulted in consumption of this product. Formic acid was also produced and subsequently consumed in some fermentations. One curious result was the accumulation of ethanol shown in figure 4, at mesophilic, neutral pH (NH_4HCO_3) conditions.

3.2 Gas Production

Figure 5 illustrates the accumulation of gas production throughout the fermentation. It can be seen that the mesophilic cultures produced larger volumes of gas more quickly. The dominant component of the gas was carbon dioxide, with very little methane being produced. The CO_2 is released from both the fermentation of the sugars and through neutralization of the product acids by the carbonate buffers.

3.3 pH effects

Figure six presents the pH of the fermentations through the experiment. The ammonium buffered fermentations dropped pH rapidly as acid production initiated. This resulted in production of lactic acid, which is typical of lower pH fermentation. As the pH was raised with the addition of additional buffer, the lactic acid was then consumed and converted into other fermentation products.

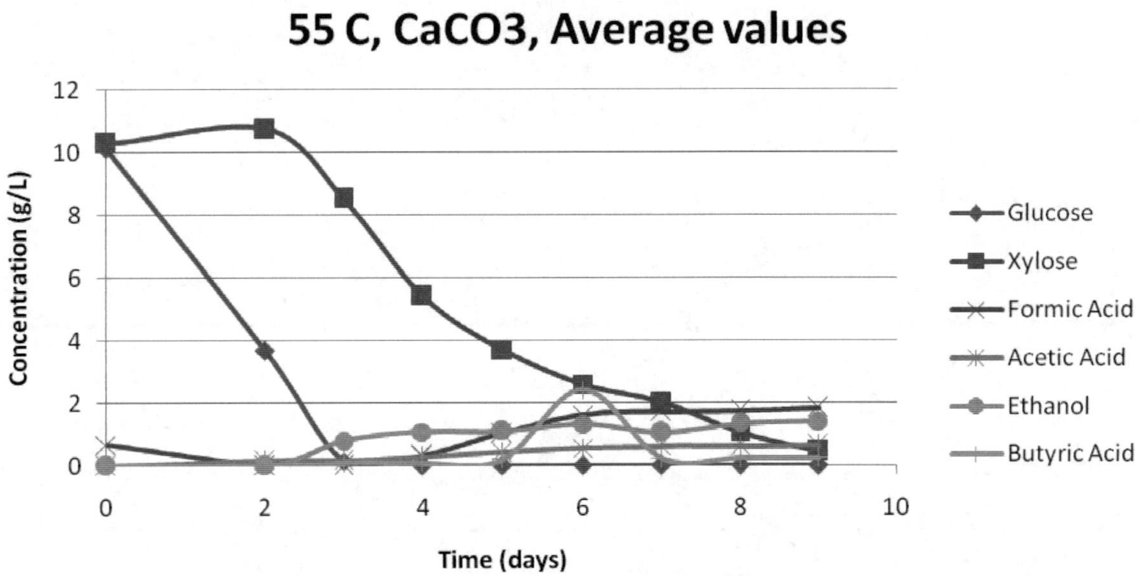

Figure 1: Fermentation results for conversion of glucose and xylose to organic acids at 55 °C with calcium carbonate as buffer. This fermentation resulted in considerable production of lactic acid (data not shown). Results shown are average values from four replicate fermentations.

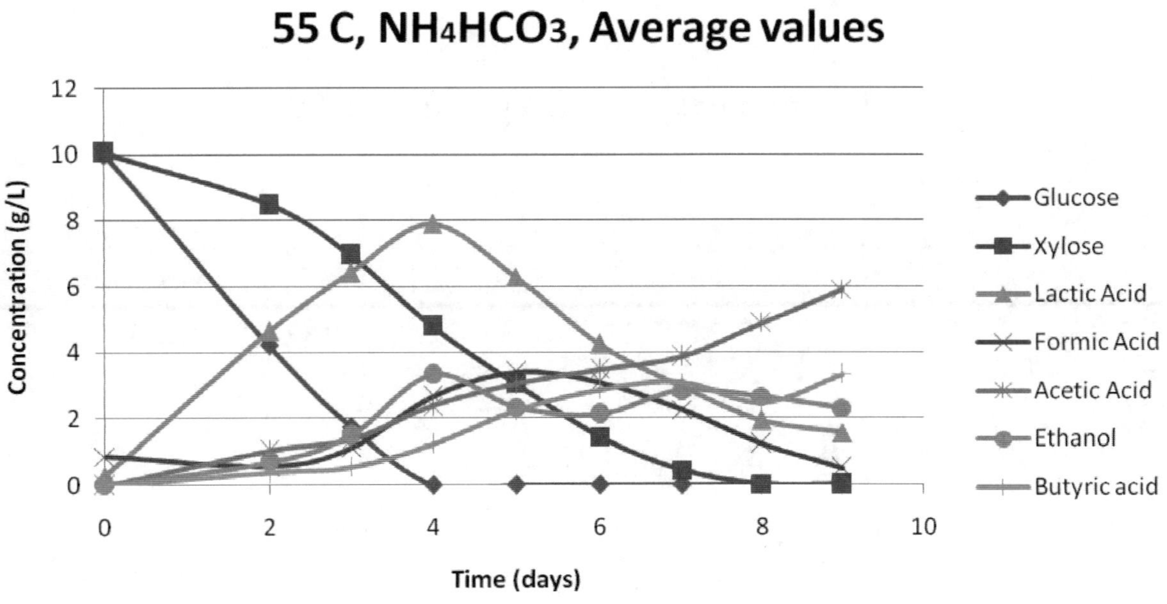

Figure 2: Fermentation results for conversion of glucose and xylose to organic acids at 55 °C with ammonium bicarbonate as buffer. In this case lactic acid accumulated and was then consumed. Results shown are average values from four replicate fermentations.

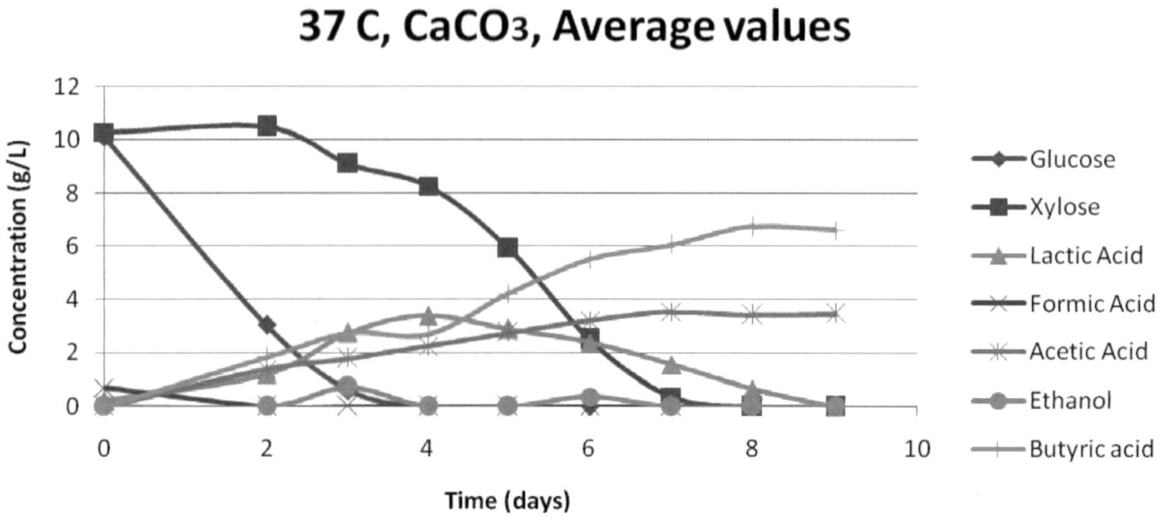

Figure 3: Fermentation results for conversion of glucose and xylose to organic acids at 37 °C with calcium carbonate as buffer. Butyric acid is the most predominant product. Lactic acid is produced and then consumed. Results shown are average values from four replicate fermentations.

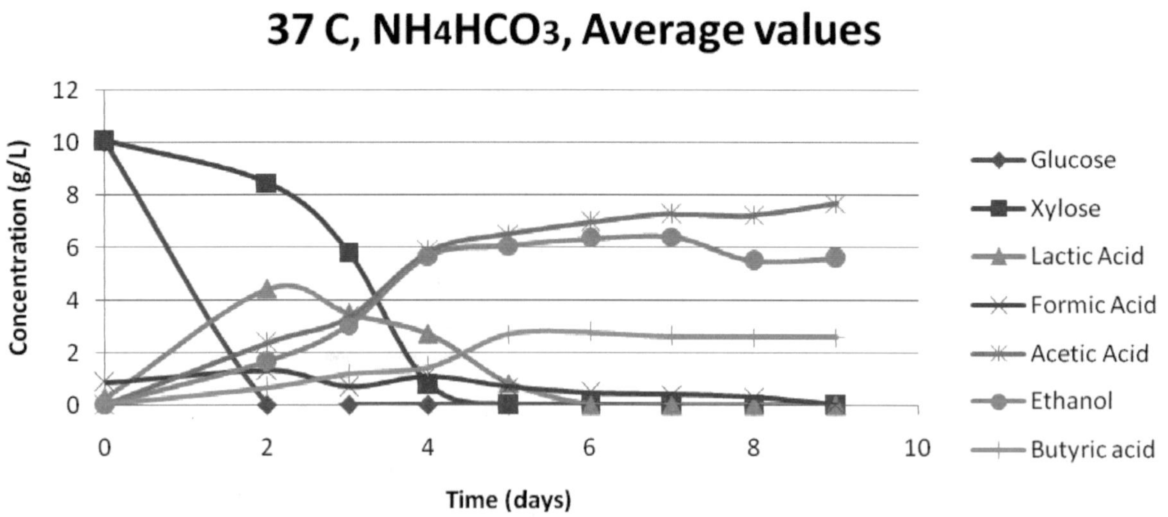

Figure 4: Fermentation results for conversion of glucose and xylose to organic acids at 37 °C with ammonium bicarbonate as buffer. Butyric acid and ethanol are both prominent products and lactic acid was produced and then consumed. Results shown are average values from four replicate fermentations.

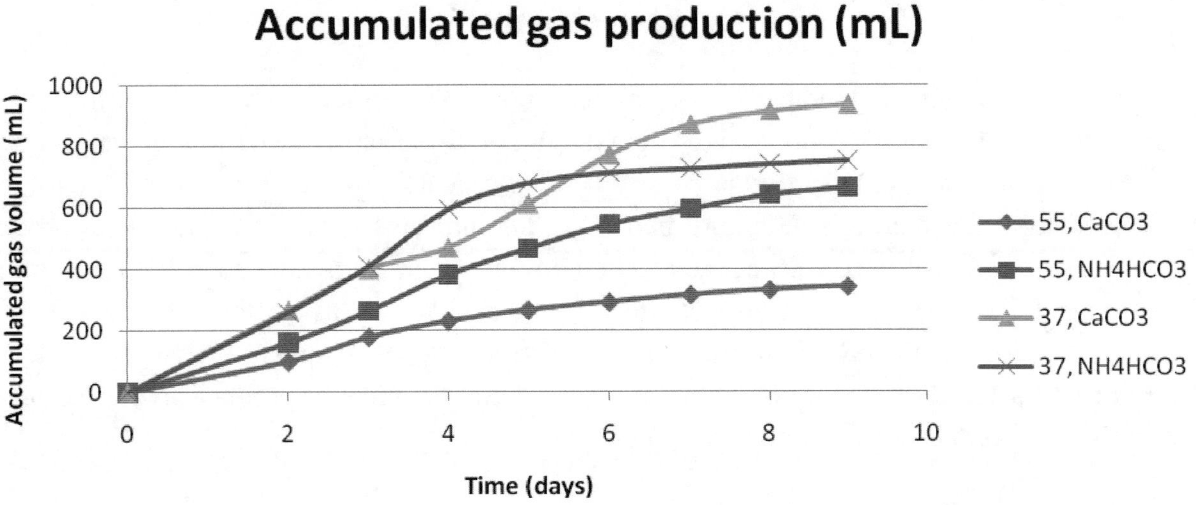

Figure 5: Gas volume production results for conversion of glucose and xylose to organic acids at 55 and 37 °C with calcium carbonate and ammonium bicarbonate as buffer. Gas composition in all cases was largely carbon dioxide with little methane detected. Mesophilic growth shown the most copious gas production.

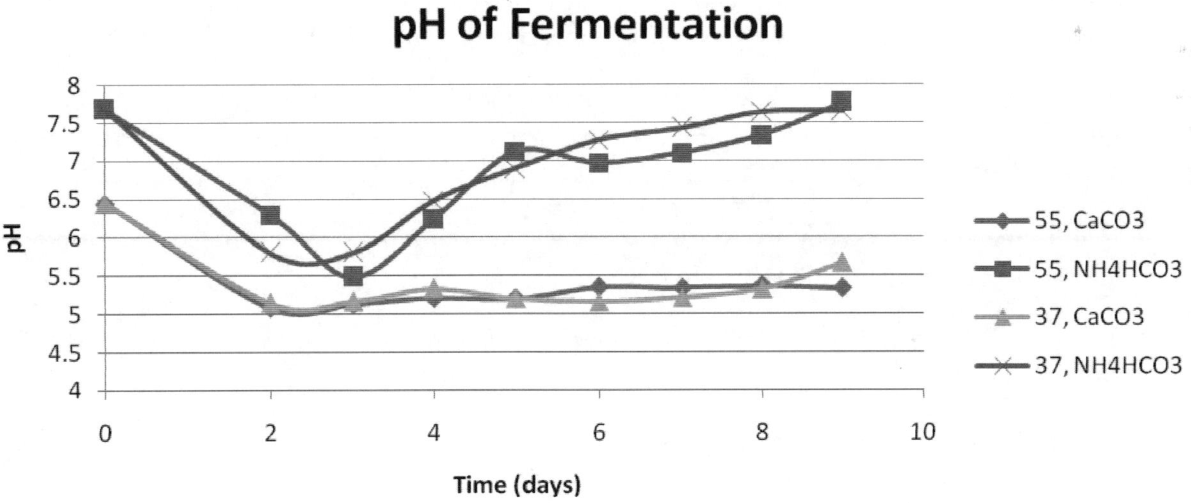

Figure 6: pH levels for conversion of glucose and xylose to organic acids at 55 and 37 °C with calcium carbonate and ammonium bicarbonate as buffer. Calcium carbonate is unable to maintain pH above 5.5, whereas ammonium bicarbonate can maintain the pH at neutral levels.

IV Conclusion

Preliminary results indicate that mixed microbial cultures derived from saline environments are capable of converting mixed sugars to organic acids. Ambient pH and temperature conditions resulted in widely variable, but repeatable product profiles. Since these were derived from identical mixed culture innocula, this indicates strong dependence of product profile on culturing conditions. Optimal acid production was achieved at mesophillic temperature and neutral pH, resulting from the use of ammonium bicarbonate as buffer. These conditions provided the most rapid conversion and highest accumulation of organic acids. The calcium carbonate as a buffer agent could not maintain a neutral pH for optimal production of carboxylic acid and thereby produced lactic acid as a byproduct.

IV References

1. George W. Huber, Sara Iborra and Avelino Corma (2006). Synthesis of Transportation Fuels from Biomass: Chemistry, Catalysts, and Engineering. Chemical Reviews, Vol. 106, No. 9

2. Joseph B. Binder† and Ronald T. Raines (2009). Simple Chemical Transformation of Lignocellulosic Biomass into Furans for Fuels and Chemicals. Journal of American Chemical Society. Vol. *131*: 1979–1985

3. Van Heningen, A. (2006) Converting a kraft pulp mill into an integrated forest products biorefinery. PulpPaper Canada 107:38-43

4. Tunc, Mehmet Sefik Sefik(2008) Hydrothermal dissolution of mixed southern hardwoods. Holzforschung. 62 :539-545.

5. Byung-Hwan Um *&* G. Peter van Walsum (2009) Acid Hydrolysis of Hemicellulose in Green Liquor
6. Pre-Pulping Extract of Mixed Northern Hardwoods Appl Biochem Biotechnol 153:127–138

7. Mark T. Holtzapple, Richard R. Davison, M. Kyle Ross, Salvador Aldrett-Lee, Murlidhar Nagwani, Chang-Ming Lee, Champion Lee, Seth Adelson, William Kaar, David Gaskin, Hiroshi Shirage, Nan-Sheng Chang, Vincent S. Chang, and Mitchell E. Loescher (1999) Biomass Conversion to Mixed Alcohol Fuels Using the MixAlco Process Applied Biochemistry and Biotechnology Vol. 77–79:609-631

8. Balaban, M. (2003) Estimation of volatile acids in wood and bark. Holz als Roh- und Werkstoff 61:465-468

9. Ross Michael Kyle(1998) Production of Acetic Acid from Waste Biomass Ph.D. Dissertation, Texas A&M University, College Station, Texas.

Catalytic Production and Upgrading of Biomass Derived Monofunctional Hydrocarbons

Elif I. Gurbuz, Chemical and Biological Engineering, University of Wisconsin - Madison, Madison, WI

Edward L. Kunkes, Chemical and Biological Engineering, University of Wisconsin - Madison, Madison, WI

Dante A. Simonetti, Department of Chemical, University of California at Berkeley, Berkeley, CA

Ryan M. West, Chemical and Biological Engineering, University of Wisconsin - Madison, Madison, WI

Juan Carlos Serrano-Ruiz, Chemical and Biological Engineering, University of Wisconsin - Madison, Madison, WI

Christian A. Gaertner, Chemical and Biological Engineering, University of Wisconsin - Madison, Madison, WI

James A. Dumesic, Chemical and Biological Engineering, University of Wisconsin - Madison, Madison, WI

We have studied the deoxygenation/ reforming of biomass derived carbohydrates to yield monofunctional hydrocarbons that can be utilized as fuels, commodity chemicals and solvents. Additionally, we have developed C-C coupling processes to upgrade these monofunctional species into fuel grade components that can be utilized in the current transportation infrastructure.

We have demonstrated the catalytic conversion of biomass-derived carbohydrates, including monosaccharides and sugar alcohols, into hydrophobic mixtures of monofunctional C4-C6 hydrocarbon species containing alcohols, ketones, heterocyclic compounds and carboxylic acids. The conversion occurs over carbon supported Pt-Re catalysts and removes more than 80% of the initial oxygen content of the sugars and polyols, yielding a spontaneously-separating organic phase. This process operates at moderate pressures (20-30 bar) and temperatures (283-523 K) and utilizes highly concentrated aqueous feeds (40-60%) of sorbitol or glucose. At 503 K and 18 bar, Pt-Re/C showed excellent stability for longer than one month time-on-stream and yielded an organic stream containing ~ 50% of the carbon found in the 60 wt% sorbitol feed. A yield of 70% of the maximum possible conversion of the carbon in sorbitol to monofunctional species was obtained, corresponding to the production of 1 kg of organic for every 4 kg of sorbitol.

Furthermore, we have studied catalytic C-C coupling processes to convert functional species (carboxylic acids, alcohols and ketones) derived from carbohydrate conversion into C7-C12 ketones, that can be converted into diesel grade alkanes via deoxygenation over solid acid supported metal catalysts such as Pt/NbOPO4. These processes include ketonization - in which two carboxylic acid molecules combine to form a linear ketone, CO2 and water, and aldol condensation/hydrogenation - in which two ketone or secondary alcohol molecules combine to form a singly-branched ketone. We have studied ketonization of the aforementioned carbohydrate derived organic species over a CeZrOx catalyst at 648-673K and found near 100% conversion of carboxylic acids into C7-C11 linear ketones. The aldol condensation/hydrogenation process occurs on bi-functional catalysts that contain acid/basic functionality as well as metal sites to dissociate hydrogen. Aldol condensation/hydrogenation was studied over a low loading (0.25 wt%) Pd/CeZrOx catalyst at 623K, and was found to convert ~60% of the condensable ketones and alcohols found in the aforementioned carbohydrate-derived mixtures into C8-C12 branched ketones.

Further investigation of the aldol condensation/hydrogenation reaction was performed by examining the reactivity of a representative ketone - 2-hexanone over Pd/CeZrOx and CeZrOx catalysts at temperatures between 573 and 673 K, and pressures of 5 to 26 bar. Reaction kinetics studies show that in addition to the expected C12 condensation product (7-methyl-5-undecaone), the CeZrOx-based catalysts produce C18 and C9 secondary species, along with light alkanes (<C7). Low loadings of Pd (e.g., 0.25 wt. %) lead to optimal activity and selectivity for the production of C12 species. The high activation energy of C9 formation

(140 kJ/mol) compared to the formation of C12 and C18 species (15 and 28 kJ/mol, respectively) indicate that these species may be formed as a result of the decomposition of heavier condensation products. The self-coupling of 2-hexanone was found to be positive order in both 2-hexanone and hydrogen. The addition of primary alcohols and carboxylic acids as well as water and CO2 to the feed was found to reversibly inhibit the self-coupling activity of 2-hexanone.

See more of this Session: Production of Fungible Biofuels From Lignocellulose
See more of this Group/Topical: Forest and Plant Bioproducts Division

ENERKEM'S THERMOCHEMICAL BIOREFINERIES: Converting Heterogeneous Biomass Into Alcohols

David Lynch, Enerkem Inc., Montreal, QC, Canada

INTRODUCTION

Enerkem's proprietary thermo-chemical technology platform is designed to use non-homogeneous waste as feedstock to produce a clean syngas suitable for catalytic synthesis. This opens the door to the use of negative-cost feedstock such as municipal solid waste (MSW) and contaminated biomass such as de-commissioned electricity poles. Enerkem can produce many valuable transportation fuels and green chemicals using a wide variety of feedstock due its extensive gas conditioning process which produces a tailored syngas that can be converted into alcohols using established and readily available industrial catalysts at the proper temperatures and pressures. Enerkem's products include methanol and ethanol, other advanced fuels, such as synthetic diesel, dimethyl ether, and synthetic gasoline, as well as other green chemicals. Additionally, Enerkem's gasification technology allows it to produce renewable electricity. The end-product mix makes it possible for Enerkem to develop a wide range of projects based on market conditions from biorefineries to renewable electricity plants.

Enerkem's proprietary technology platform has been developed completely in-house. Since 2000, the Company has followed a rigorous path to support commercial viability and proceeded through each developmental step: lab to pilot to demo commercial plant. The incremental steps in commercialization resulted in more than 3,600 hours of exhaustive piloting and optimizing of both design and feedstock processing.

With its multi-product and feedstock-flexible technology platform, Enerkem helps reduce greenhouse gas emissions, while contributing to a greener economy. By producing fuels locally rather than depending on foreign oil, Enerkem contributes to increasing energy independence and reducing imports.

OVERVIEW

The development of Enerkem's technology platform has been done completely in-house. R&D work started in 1999 with the bench scale syngas island. Over 1,000 testing hours were conducted. Twelve R&D staff was devoted to the work. At that time, the Company started to patent its technology.

In 2003, the Company started operating its pilot plant in Sherbrooke, Québec. The pilot plant produces syngas, methanol, and second-generation ethanol, and is equipped with various sampling ports for data collection. It is also well instrumented and is automated for testing and reporting. The system can feed solid materials, slurries, and liquids. To date, approximately 20 different feedstocks have been used to test and validate the technology, and for engineering design purposes. These feedstocks include:

• Municipal Solid Waste (4 cities);

• Forest biomass (as well as treated wood and C & D wood, both pelletized and chipped);

• Agriculture biomass (wheat straw, corn stovers);

• Plastics (various forms: pelletized, shredded, fluffed and mixed);

• Glycerine (rich fraction of biodiesel production after transesterification);

• Biosolids (dried, from Montreal and from pulp & paper operation);

• Petcoke (Alberta, with 8% humidity and 5-6% sulfur)

Unlike many project developers that license outside technology, Enerkem's technical team's intimate knowledge of the technology and its dedication to the technology development efforts give it the necessary internal capacity to engineer, build and operate its own plants.

Enerkem's first commercial-scale biofuels and biochemicals plant is currently producing its premium syngas. It is located in Westbury, Quebec. It will be the world's first ethanol plant to use negative-cost and unconventional materials - treated wood from used electricity poles. The plant will, at term, produce 1.3 million gallons of second-generation ethanol annually.

The Company will also build and operate a plant that will produce 10 million gallons of next generation biofuels annually, from sorted municipal solid waste. The City of Edmonton will supply a minimum of 100,000 tonnes of sorted municipal solid waste per year. Construction is expected to start in 2009. Another commercial project is under development. The Company will transform sorted municipal solid waste and woody biomass in Pontotoc, Mississippi. It plans to produce 20 million gallons of ethanol per year.

In Enerkem's bubbling fluidized-bed gasifier, a continuous fee of biomass is carried into a reactor where an inert heat carrier (i.e. sand) is "fluidized" in a zone known as "the bed" under relatively low temperatures and moderate pressures, hence called "low severity gasification". Enerkem's technology platform consists of three distinct process steps: gasification of biomass; cleaning and reforming of synthetic gas (syngas), and; conversion of syngas into biofuels and green chemicals via catalysis. In Enerkem's bubbling fluidized-bed gasifier, a continuous feed of biomass is carried into a reactor where an inert heat carrier (i.e., sand) is "fluidized" in a zone known as “the bed” under relatively low temperatures (700–750°C) and moderate pressures (~2 atm.). In this environment, biomass is converted into gas and small amounts of ash. Oxygen and steam are used as fluidizing gases and gasification agents. Gasification takes place throughout the bed. The injected oxygen and added steam prevent the ash from melting and maintain the injection nozzles at appropriate temperatures, so available steels can be used, thus avoiding exotic and expensive ceramics. The syngas is drawn from the top of the gasifier. In its movement towards the top of the bed, the synthetic gas disengages from the sand and undergoes extensive cracking and reforming in the freeboard stage – key chemical processes, which convert organic intermediates to simple molecules, having the desired molecular characteristics (H_2, CO, CH_4 and CO_2 – the main constitutive molecules in synthetic gas). The superior mixing, which occurs in the bubbling fluidized bed used by Enerkem, generates high rates of heat and mass transfer which subsequently yield stable temperatures, resulting in well controlled reaction kinetics. Thus, Enerkem's reactor is capable of handling feedstock that is not uniform, is geometrically dissimilar and is heterogeneous. Enerkem's reactor is also efficiently sized due to high throughput rates facilitated by the design – a feature that contributes to the commercial viability of the Enerkem platform.

The raw syngas produced in Enerkem's gasifier is subjected to sequential conditioning as follows: (i) cyclones remove particulates (pilot plant capture efficiencies are ~90%); (ii) secondary cracking and steam-carbon reactions reduce tar and carbon loads for the next step, catalytic steam reforming, which is carried out in a fluid bed configuration, using proprietary reforming catalysts; (iii) heat recovery is performed on the reformed syngas, which is quenched and scrubbed through a two stage system, that ensures removal of any residual Cl and S and that recovers the NH_3 produced. The latter is recycled back or is separated and sold; (iv) any residual tar or light hydrocarbon that has escaped the fluid bed reforming operation is recovered and, as an emulsion, reinjected into the gasifier where it is converted into syngas, as well as green chemicals, including acetic acid and acetates.

The catalytic conversion into alcohols is the fourth step of the process. The cleaned syngas is a mixture of H_2, CO, CO_2, residual argon and N_2 and residual methane (only traces of other light hydrocarbons are present in the syngas). The molar ratio H_2/CO resulting from the reforming is essentially 2 with no necessity of shift conversion. Should a specific molar ratio be desired, the reforming step can be adjusted, or a classical shift conversion can simply be added, either before or after the compression step. Enerkem's extensive gas conditioning process produces a tailored syngas that can then be converted into alcohols using established and readily available industrial catalysts at the proper temperatures and pressures. Enerkem can produce methanol and ethanol, other advanced fuels, such as synthetic diesel, dimethyl ether, and synthetic gasoline, as well as other green chemicals.

Enerkem has developed a modularized compact design of “100,000 metric tons in - 10 million gallons of EtOH out”. Its manufacturing system is easily scalable with skidded 10MM gallon-per-year reaction vessels that can be used in an array of larger plants. This gives the company more flexibility in engineering plants of different sizes and allowing it to replicate the design without significant re-engineering.

The use of waste biomass as feedstock is economically attractive given the negative cost of the raw material. However, the presence of contaminants in the waste requires an effective sequence of conditioning steps in the overall process, to: (a) produce a syngas that is clean and; (b) handle contaminants in a manner that they become fully neutralized and thus can be safely disposed of.

Gas- and Aqueous-Phase Hydrogenation of Acetic Acid Over Ru Catalysts

George W. Huber, Chemical Engineering, University of Massachusetts, Amherst, MA

Hakan Olcay, Chemical Engineering, University of Massachusetts, Amherst, MA

Ye Xu, Center for Nanophase Materials Sciences, Oak Ridge National Laboratory, Oak Ridge, TN

Research on biofuel production from biomass has rapidly increased due to the depletion in fossil fuel reserves and to the rising concerns with greenhouse gas emissions. Hydrogenation reactions, and particularly the hydrogenation of organic acids, are critical in most pathways for production of biofuels. This work presents and compares the results of Ru catalyzed acetic acid hydrogenation in both gas and liquid phases in various types of reactor configurations. In this presentation we will show that the aqueous phase does not change the catalytic properties, and that the catalytic activity is a function of the concentration of both the acetic acid and the hydrogen in the reactor. We will compare our experimental results with density functional theory (DFT) studies to gain a better understanding of the reaction chemistry.

All experiments were carried out with a Ru/Carbon catalyst. We studied the effect of water concentration at 175"aC in a flow reactor by keeping the acetic acid and hydrogen concentrations constant (0.0016 M and 0.061 M, respectively), and by varying the water/water vapor concentration in the system (0-0.13 M). We have observed that the presence of liquid water has very little effect on the activity and selectivity of the catalyst. It has been concluded that the hydrogen and acetic acid concentration on the catalyst surface changes the catalytic properties.

We also compared acetic acid hydrogenation in both batch and flow reactors. Again in these experiments the hydrogen and acetic acid partial pressures are the controlling variables. In the flow reactor the hydrogen partial pressure is typically significantly higher than in typical batch reactors. Therefore, flow reactors are significantly better hydrogenation reactors than batch reactors.

These results provide important insights as to how hydrogenation reactions can be improved.

See more of this Session: Production of Fungible Biofuels From Lignocellulose
See more of this Group/Topical: Forest and Plant Bioproducts Division

Commercializing Pyrolysis Oil Into Motor Fuels Leveraging the Existing Refining Infrastructure

Thomas D. Foust, National Renewable Energy Laboratory, Golden, CO

Robert Baldwin, National Bioenergy Center, National Renewable Energy Laboratory, Golden, CO

Jim Frederick, National Bioenergy Center, National Renewable Energy Laboratory, Golden, CO

Fast pyrolysis of biomass shows considerable promise as a conversion route for converting lignocellulosic biomass into fungible motor fuels. Unlike other biomass conversion routes that produce a gaseous or sugar intermediate, pyrolysis produces a liquid intermediate that could potentially be processed at an existing petroleum refinery into conventional motor fuels synergistic with conventional crude oil refining. Processing pyrolysis oil at conventional refineries would significantly reduce costs of processing and facilitate the rapid transition into the market place. However, raw pyrolysis oil has numerous qualities which will prevent it from being processed at conventional refineries such as its corrosiveness, insolubility in hydrocarbons and thermal and temporal instability. Because of these characteristics, raw pyrolysis oils must be upgraded before they would be acceptable to enter a refinery for further processing into conventional motor fuels.

Raw pyrolysis oils typically contain up to 40% oxygen and this high oxygen content is linked to many of the undesirable properties of pyrolysis oils, hence oxygen removal is critical. A detailed market assessment and experimental program has been undertaken to address the challenges of introducing pyrolysis oils into conventional refineries from both a technical and cost barrier perspective. This study looked at various intermediate oxygen levels to determine the physical and chemical properties and the processing costs associated with achieving these intermediate states. Several scenarios were identified where an initial state processing to approximately 10% oxygen would be sufficient to allow entry into a conventional refinery for ultimate processing to conventional motor fuels at approximately 0.2% oxygen. This two stage processing maximizes the advantages of economies of scales and existing infrastructure utilization. Results of this program will be shown that indicate that this approach shows considerable promise for producing conventional motor fuels from biomass at costs highly competitive with petroleum derived fuels.

See more of this Session: Production of Fungible Biofuels From Lignocellulose
See more of this Group/Topical: Forest and Plant Bioproducts Division

Thermal Pretreatment Options for Thermochemical Conversion of Lignocellulosic Biomass

Wei Yan, Chemical Engineering, University of Nevada, Reno, Reno, NV

Charles J. Coronella, Chemical Engineering, University of Nevada, Reno, Reno, NV

Tapas Acharjee, Chemical Engineering, University of Nevada, Reno, Reno, NV

Victor R. Vásquez, Chemical Engineering, University of Nevada, Reno, Reno, NV

Thermochemical conversion of lignocellulosic biomass is a promising technology for production of renewable power, fuels, and chemicals. Both pyrolysis and gasification of biomass have significant technical barriers that must be eliminated for successful widespread commercialization. Feedstock handling is complex due to the diverse nature of important feedstocks, including commercial timber, slash removed due to fire hazard reduction, switchgrass, and agricultural residues as diverse as rice hulls, corn stover and wheat chaff. Seasonal availability and low forest density make feedstock logistics complex and expensive. A process to homogenize the feedstocks that will simultaneously produce a stable energy-dense fuel is needed. Two such processes are considered here: dry torrefaction, and wet torrefaction.

Dry torrefaction is a low-temperature pyrolysis process in which the biomass is heated in a chemically inert environment at temperatures ranging between 200 and 300 °C. The reaction produces two products, a solid and a gas. The solid has about 70% - 80% of the mass, and approximately 90% of the fuel value of the original biomass.

Biomass is treated in hot compressed water in the case of wet torrefaction, also known as hydrothermal pretreatment. This process results in three products, including gases, aqueous chemicals, and a solid fuel. The temperature is in the range of 225° - 275 °C, and the pressures are up to 60 bar. The solid product contains about 50 - 70% of the mass and 70 - 90% of the fuel value of the original feedstock. The gas product is about 10%, and the aqueous chemicals make up the balance.

We have characterized experimentally the solid fuels resulting from both processes, which indicate significant similarity. Solid product from both processes have an enhanced energy density relative to the starting feedstock, and are is friable (and might be easily pelletized.) In both cases, an ultimate analysis of the solid fuels reveals decreased oxygen content and increased carbon content. In both cases, a fiber analysis of the fuel demonstrates that a significant fraction of the original hemicellulose component has been extracted. Proximate analyses reveal fuels suitable for gasification, and possibly for pyrolysis. The solid fuel in both cases exhibits hydrophobic qualities, as demonstrated by equilibrium moisture.

See more of this Session: Production of Fungible Biofuels From Lignocellulose
See more of this Group/Topical: Forest and Plant Bioproducts Division

Pyrolysis investigations and ash melting behaviour of agricultural residues used as a fuel in small furnace installations

Martina Poppenwimmer[1], *Harald Raupenstrauch*[1]

[1]*Chair of Thermal Processing Technology, University of Leoben, Franz-Josef-Strasse 18, 8700 Leoben, Austria (martina.poppenwimmer@unileoben.ac.at).*

Abstract

During thermal combustion of solid fuel in furnace installations the applied solid is first dried, then pyrolysed and finally the gasification and combustion reactions take place. In this process the pyrolysis is an important sub-step, in which the volatiles are released. Especially in case of biomass it is of special interest, since more than 85 % of the solid may be converted.

Unfortunately biomass like agricultural residues involves various problems when used as fuel in small furnace installations. The most important reasons are high ash contents on the one hand side and the low melting points of ash on the other, which are mainly caused by a high rate of alkali metals. For preventing the agglomeration of ash it is necessary to provide ideal combustion conditions and to avoid temperature peaks. Therefore it is essential to get more detailed information of devolatilisation behaviour of the various fuels.

In addition to furnace design and operation conditions it is possible to manipulate the fuel properties by additives in order to reduce ash melting and agglomeration. Concerning the selection of additives attention was paid to the high content of alkaline earth metals, because these elements increase the ash melting point.

The results of the tests show on the one hand the influence of several additives to ash melting behaviour and on the other the effect on the progression of calorific value of formed volatiles.

Introduction

Because of still decreasing fossil reserves it is consequently searched for alternative fuels. One example is wood, but it is essential to minimize the dependence on wood industry. Therefore other biomass like agricultural residues is analysed. However there are some difficulties and problems in use of these residues for generation of energy. In addition to the high ash content and the low ash melting point also the nitrogen content of the fuel is a problem. But it is possible to reduce it fractionally by means of elution. Due to washing of the residues also the unwanted ingredients chlorine and potassium can be reduced and furthermore corrosion damages by condensation on cold plant components can be avoided. However the washing of straw involves high costs.

Concerning the low melting behaviour of ash it is also caused by fuel content (see Fig.1).

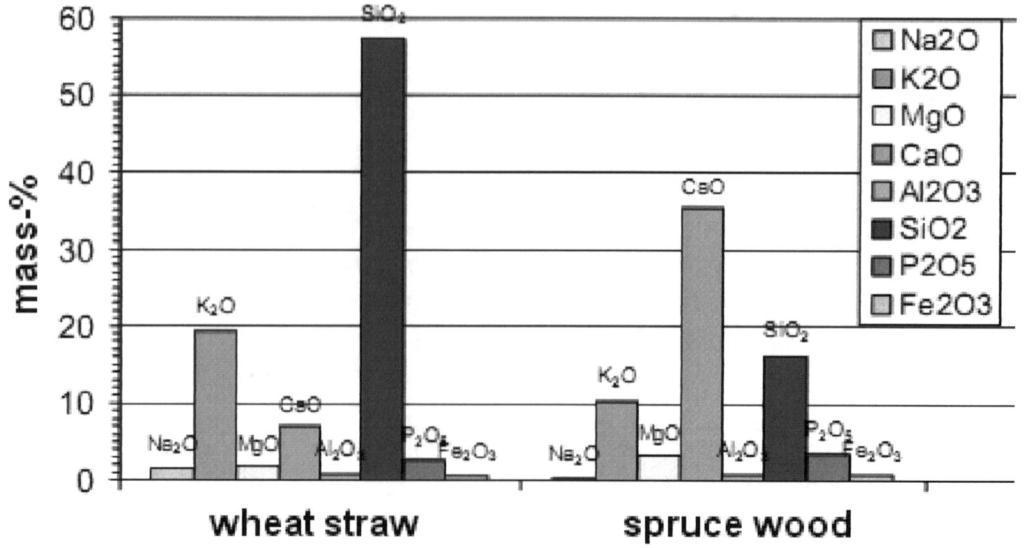

Figure 1: Typical values for ash composition [1]

Compared with spruce wood wheat straw contains high value of SiO_2 and K_2O. Alkaline metals like potassium and sodium lower the ash melting point as well as silicon. On the other hand alkaline earth metals increase the melting point. The content of these metals in wheat straw is rather low. For improving the ash melting behaviour and reducing agglomeration it is possible to manipulate the fuel properties by additives. Concerning the selection of additives attention is paid to the high content of alkaline earth metals. Therefore materials like lime flour, which includes mainly calcium carbonate with a small share of magnesium carbonate, are used. Two other possibilities, which were applied for analyses, are blast furnace slag and molasses. However, it is important to add only a smidgen of an inert substance, because the ash volume and the abrasion resistance get worse with it. Only in case of molasses the abrasion resistance might be possibly improved. [1,2,3]

Additives influence not only the melting behaviour but also the devolatilisation of the solid fuel. The analyses of pyrolysis are operated with an analysis device, which includes a thermo gravimetric scale (TGS) in combination with a differential scanning calorimeter (DSC). Because of the DSC it is possible to investigate the time dependency of the heat of reaction of volatiles and in combination with the measured mass loss the calorific value respectively. So the devolatilisation behaviour of pure wheat straw pellets was compared with them of the mixtures with additives.
With the knowledge of the exact curve progression of calorific value the yield of useable heat can be improved and emissions minimized and consequently low environment pollution will be achieved.

Analysis methods

For analysing the fusibility of fuel ash, first the biomass residues are milled to analytical grade. Then they are incinerated in a muffle oven according to CEN TC335/WG N43 (standard for investigation of ash content of solid biomass fuels). Finally the ash is detected with a LEITZ hot stage microscope according to ÖNORM CEN/TS 15404 (Solid derived fuels - Process for determination of fusibility of fuel ash in using various temperatures). During the investigations the sintering and softening temperature are detected. At sintering temperature the sample reaches only 95 percent of his input size. At this temperature the sintering process of the ash is starting and the particles adhere together. First indications of ash melting occur at the softening temperature.

Pyrolysis investigations are made with the pilot plant of the Chair of Thermal Processing Technology (TPT) at the University of Leoben, which combines a Thermo Gravimetric Scale (TGS) and a Differential Scanning Calorimeter (DSC). A flow sheet of this test facility is given in Figure 2.

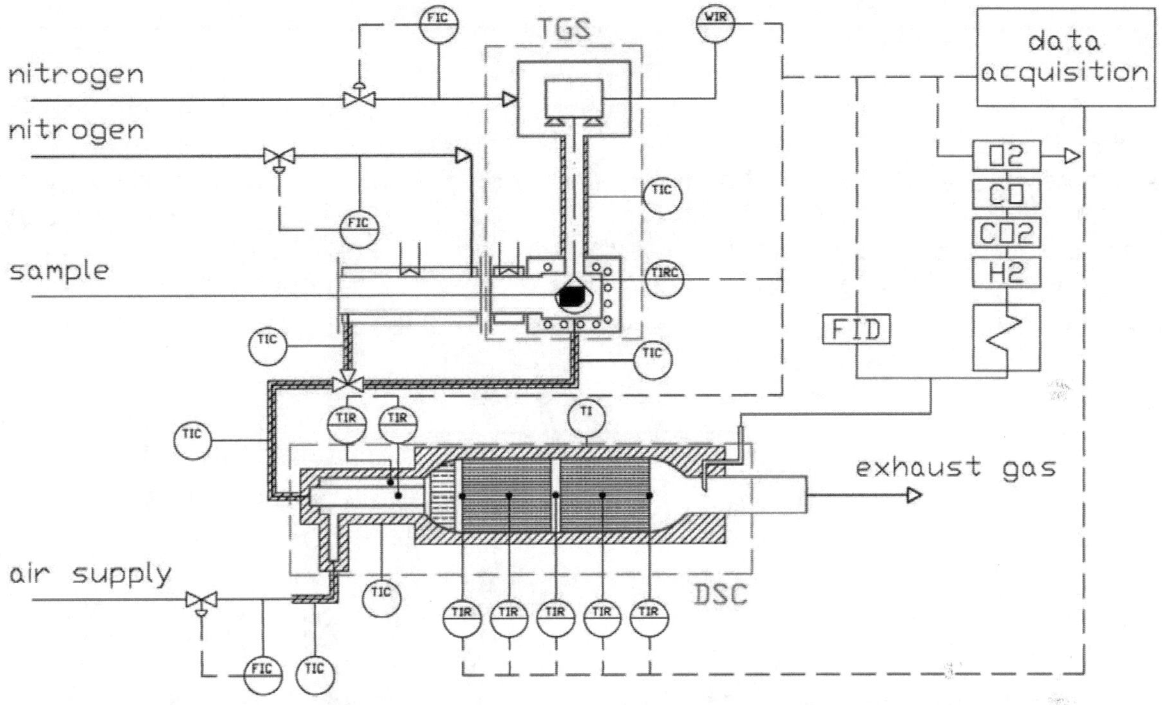

Figure 2: Flow sheet of test facility [4,5]

The fuel is pyrolysed in a muffle oven on a scale basket under nitrogen atmosphere and afterwards the produced pyrolysis gas is mixed with air. In the heat exchanger the pyrolysis gas and the combustion air are equalised and after a mixing comb burned on the surface of a platinum comb catalyst. The occurring difference of temperature in the catalyst during the combustion is quantified and with it the heat of reaction of volatiles is determined.

After the catalyst a sampling unit is implemented, which leads the flue gas to the analysers. For synthesis analyses O_2, CO_2 and CO analysers as well as a flame ionisation detector are installed.

In addition, the calorific value of the synthesis gas is also calculated, simultaneously measuring the mass loss of biomass (cf. 1).

$$H_{u(t)} = \frac{Q_{(t)}}{dm/dt} \qquad (1)$$

For determining the oxygen demand of the volatiles dependent on time, an oxygen mass balance of the DSC was performed. Thereby the oxygen entry is known because of the defined combustion air and the oxygen in the flue gas is measured. The oxygen demand of all volatiles was considered as the difference between these two values (cf. 2).

$$O_{2,dem} = O_{2,in} - (O_{2,out} + 5 \cdot C_3H_{8,eq}) \qquad (2)$$

Additionally the calorific value of the fuel was investigated with a bomb calorimeter in order to close the energy balance of the test evaluation and to get a verification of the established results compared with the theoretical ones. [4,5]

Results

Ash melting behaviour

For the investigations of ash melting behaviour the additives lime flour, molasses and blast furnace slag (BFS) were used.

The results at figure 3 show, that a larger quantity of additive does not automatically mean a further increasing of temperature. The non expectable reduced temperature can be explained with occurring eutectics. Especially in case of addition of lime flour a better increasing of the sintering temperature is noticeable with 3% additive than with 4%. The same way show the mixtures with molasses. 10% molasses involve a better effect on sintering and softening temperature than 20% molasses.

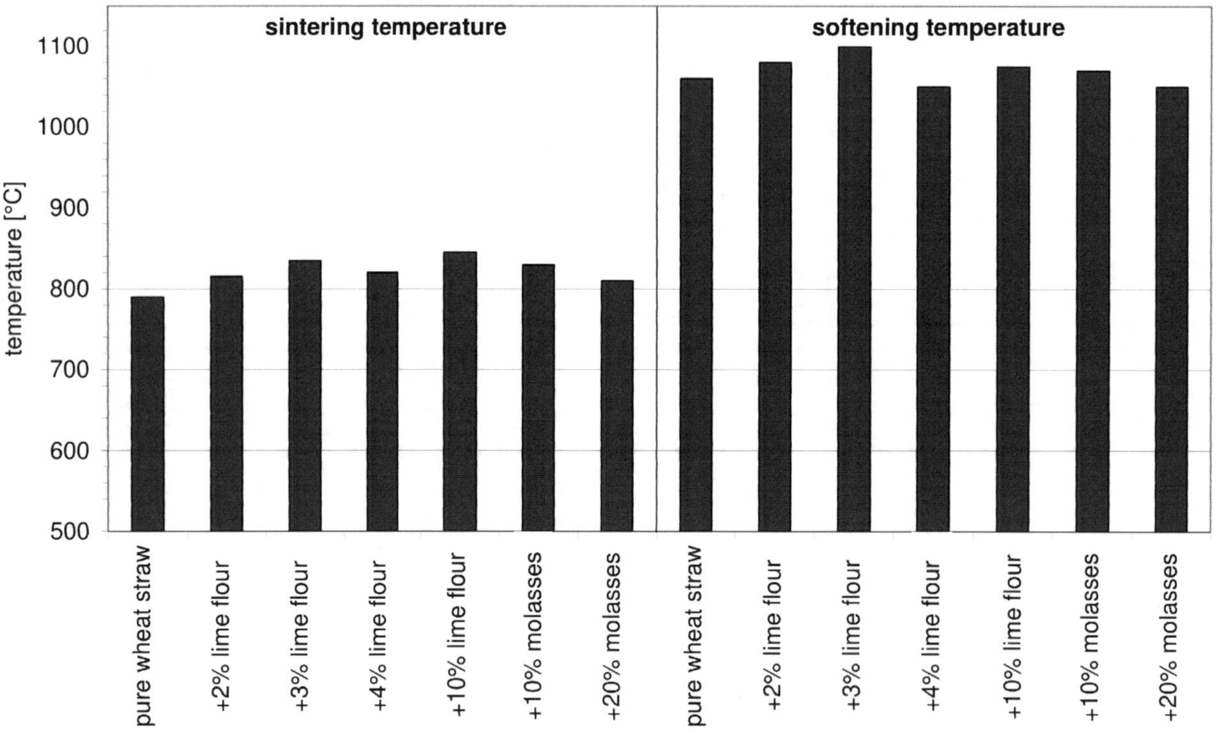

Figure 3: Comparison of the increasing of sintering and softening temperature with additives

The softening temperature of the analysed wheat straw is 1060°C. This value is rather high compared with values from literature. There you will find softening temperatures of wheat straw from 860 to 910°C.

Pyrolysis investigations

Pure wheat straw pellets and in combination with additives like blast furnace slag and molasses were determined about their devolatilisation behaviour at a temperature of 800°C.
For closing the energy balance of the evaluation, the calorific value of pyrolysis coke of fuels is investigated with a bomb calorimeter in addition to the calorific value of the solid.
Always two pellets with a length of 15 millimetres and a diameter of 6 millimetres, which weight is about 1 gram, were put on the scale basket and pyrolysed.

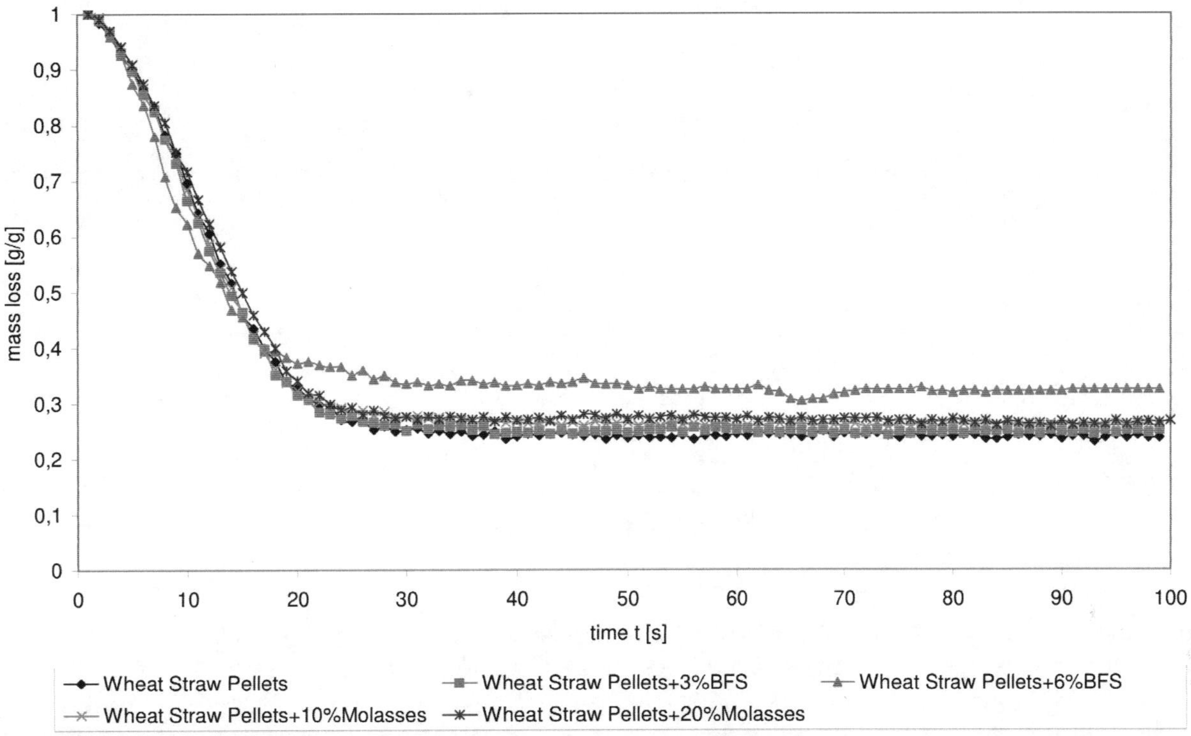

Figure 4: Mass loss of 15mm long wheat straw pellets (m_0 ~1 g) dependent on time (T_{oven} = 800°C, nitrogen atmosphere, $T_{catalyst,start}$ = 480°C)

In figure 4 it is well noticeable that the mass losses of the various samples proceed a similar way. All curves continue linear after about 30 seconds. At this moment the pyrolysis of the fuel is finished. The small variations at the end of the mass curves are effected on the one hand by the vibrations of the scale, which are caused by the perfusion of the furnace chamber and on the other hand probably by the small sample mass, whereby the solid fuel can't stand smooth on the basket cage. But these variations do not influence the following evaluations, because the mass loss curve is fitted by a proper function (cf. 3).

$$m_{loss} = \frac{a + c \cdot t^2 + e \cdot t^4}{1 + b \cdot x^2 + d \cdot x^4} \qquad (3)$$

Considering figure 4 it is observable that the share in pyrolysis coke is different for each sample. The lowest value is measured for pure wheat straw pellets (23%). This means that 77 % of the solid were pyrolysed. For wheat straw with 6% blast furnace slag 32% coke are detected. Because of the inert blast furnace slag the content of ash is much higher than in case of molasses as additive and following also the coke fraction is the biggest. Coke is composed of ash and fixed carbon.

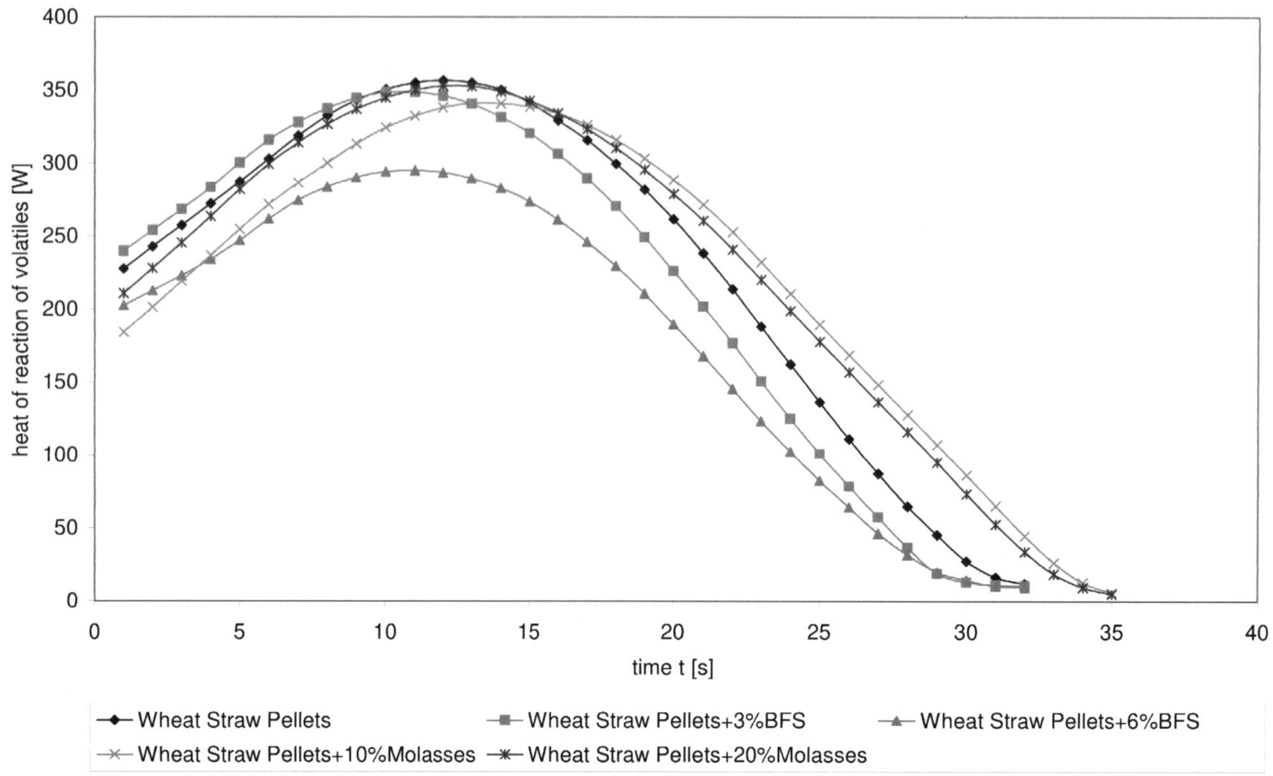

Figure 5: Heat of reaction of all volatiles of 15mm long wheat straw pellets (m_0 ~1 g) dependent on time (T_{oven} = 800°C, nitrogen atmosphere, $T_{catalyst,start}$ = 480°C)

Figure 5 shows the average values of heat of reaction of volatiles from all experiments. The curve of pure wheat straw pellets has got the highest maximum and the total released energy the highest value as well. Due to addition of inert blast furnace slag the maximum and also the total energy decrease. This confirms the results of the measurements with bomb calorimeter. The sum of the heat of reaction of volatiles from wheat straw pellets with 10% molasses and 20% molasses is nearly the same as the one of pure wheat straw. Molasses is unlike blast furnace slag an organic material, which has got a calorific value a little lower than the one of wheat straw.

The courses of all heat of reactions of volatiles proceed in a similar way. At the beginning many volatiles release, because the energy increases rapidly. The end of pyrolysis is given after about 30 seconds for the experiments with blast furnace slag as additive. The other pyrolysis processes of wheat straw with molasses end approximately after 35 seconds. This can be explained with structure of the pellets. Pellets in combination with blast furnace slag are not so compact like those with molasses, which acts as a binding material and thereby volatiles can not exhaust so easily.

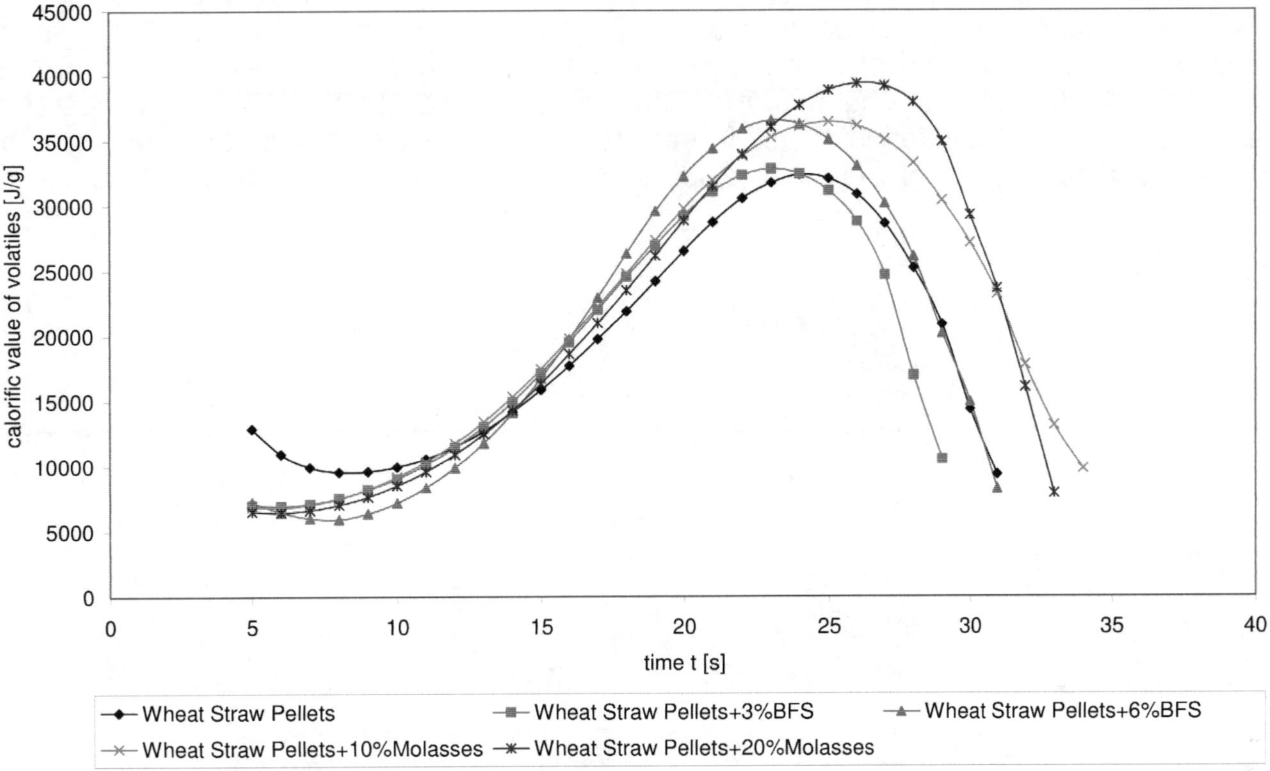

Figure 6: Calorific value of volatiles of 15mm long wheat straw pellets (m_0 ~1 g) dependent on time (T_{oven} = 800°C, nitrogen atmosphere, $T_{catalyst,start}$ = 480°C)

The calorific value of the volatiles (cf. figure 6) rises to a maximum at the end of pyrolysis. At the beginning only volatiles with a low specific calorific value exhaust. The end of the averaged devolatilisation of each sample combination differs a little bit. First the calorific value curve of wheat straw with 3% blast furnace slag ends after about 30 seconds. Second finish the one of pure wheat straw and 6% blast furnace slag. The two with molasses addition have got the longest exhausting time, which could be seen as well in figure 5. The compacter construction announces furthermore the high maxima of the two curves with molasses. Because of this additive the pores at the outside of the pellet are closed and the volatiles have to outgas axial. So the longer condensable hydrocarbons (tar) are easier to split on the surface of the hot ash-coke construction into shorter compounds with a high rate of hydrogen.
Blast furnace slag has also got a positive effect on cracking reactions and degassing of components with a high calorific value.

Conclusions

The results of the ash melting experiments showed that it is possible to raise the sintering and softening temperatures with addition of suitable substances. However the increasing was not very high with small amount of additives. But a larger quantity would worsen the abrasion resistance and the ash volume. Certainly it is also necessary to enhance the furnace installations for avoiding temperature tops and following ash agglomerations.
Therefore the information of pyrolysis investigation can be applied. Concerning the devolatilisation, all samples showed a similar characteristic. At the beginning the heat of reaction increased rapidly, just as the mass loss fell off. Also the length of pyrolysis was nearly the same except for the samples with molasses, which lasted a bit longer. Molasses

closed the outer pores and supported the axial emission of pyrolysis gases. Due to this fact also the characteristics of calorific value of volatiles differed from the samples without additives. Towards the end of pyrolysis the specific calorific value increased higher. This points to a better tar cracking tendency into shorter components with a high share in hydrogen. Also the addition of blast furnace slag showed a rising of the maximum of calorific value curve and consequently a possible positive effect on cracking reactions.

Table of symbols

SiO_2	silicon oxide
K_2O	potassium oxide
O2	oxygen
H_u	calorific value [J/g]
m	masse [g]
Q	energy [J]
t	time [s]
dem	demand
eq	equivalent

References

1 Ottmann M., Verbrennung biogener Brennstoffe in stationären Wirbelschichtfeuerungen. PhD thesis, München University of Technology, 2006

2 Bensmann M., Stroh vergolden. Neue Energie - Erneuerbare Energie/ Biomass, Volume 03, 2006

3 Kiesewalter S. and Röhricht Ch., Pelletproduktion aus halmgutartiger Biomasse. Waemre 2004, 2004

4 Poppenwimmer M., Reaktionswärmestrom von Flüchtigen und Teer in Abhängigkeit der Zeit bei der Pyrolyse von festen Brennstoffen. Master's thesis, Graz University of Technology, 2007

5 Maurer Ch., Development of a Differential Scanning Calorimeter - Experimental Investigations of Wood and Plastics. PhD thesis, Graz University of Technology, 2006

Dissolution and Fractionation of Wood and Straw Using Ionic Liquids

Miguel A. Tavares Cardoso, Delft ChemTech, Delft University of Technology, Delft, Netherlands

Gianluca Marcotulio, Process & Energy Laboratory, Delft University of Technology, Delft, Netherlands

Jaap van Spronsen, Process & Energy Laboratory, Delft University of Technology, Delft, Netherlands

Geert-Jan Witkamp, Process & Energy Laboratory, Delft University of Technology, Delft, Netherlands

W. de Jong, Process & Energy Laboratory, Delft University of Technology, Delft, Netherlands

J. Ruud Van Ommen, Delft ChemTech, Delft University of Technology, Delft, Netherlands

The decrease in the resources of fossil fuels together with the global warming alarms have led to an intensified search for alternative fuels to supply the modern society. Among them, the concept of biorefinery is becoming increasingly important since it presents the potential to be a real alternative for large scale production of fuels with a neutral carbon dioxide balance. In this biorefinery context, biomass resources such as wood and straw have not yet been properly valorised since these materials are mostly incinerated to produce heat. In addition, production of fuels from lignocellulosic biomass, especially wood, minimizes the competition with the food chain and presents higher yields when compared to first generation biofuels. The challenge now is to make this process attractive from an economical point-of-view. One of the major difficulties when dealing with these raw materials is related with the need of pre-treatments to dissociate the biomass structure. Lignocellulose consists mainly of cellulose (35 -50%), hemicellulose (20 -35%) and lignin (5-30%). These components are assembled in a complex three-dimensional structure remarkably resistant against chemicals and microbial attacks that makes them very difficult to hydrolyze, which is key for its future utilization. An effective dissociation of these components and their separation can lead to the production of several high value products prior to the production of fuels. This integrated approach increases the competitiveness of the biorefinery industry and can make it attractive even considering the actual oil prices (2009). As previously reported, ionic liquids offer a new and attractive route for the dissolution of lignocellulosic biomass. Besides the amount of biomass that can be dissolved in some ionic liquids (that can reach 8 % w/w), their complete recycling and significant environmental advantages make this class of compounds the best possible choice for a new pre-treatment to dissolve and fractionating the lignocellulosic biomass prior to biorefinery processing.

In this work, the complete dissolution of both wood and wheat straw in ionic liquids was followed by the fractionation of the several components that constitute the biomass and the analysis of each fraction was performed to close the mass balance. Two different ionic liquids, 1-ethyl-3-methylimidazolium chloride and 1-butyl-3-methylimidazolium chloride, and their mixtures were tested. The results were compared both in terms of yield and selectivity for the different constituents of both types of biomass. The obtained results show that this is a promising process for the pre-treatment of lignocellulosic biomass, regardless the type (wood or straw). Moreover, since the ionic liquids can be almost 100% recycled, this is a low cost, fast, simple and environmentally friendly pre-treatment that can increase significantly the competitiveness of modern biorefineries.

See more of this Session: Integrated Biorefineries for Lignocellulosic Biomass
See more of this Group/Topical: Forest and Plant Bioproducts Division

An Assessment of The Recovery and Recycle of Ionic Liquids (ILs) Following Lignocellulosic Biomass Pretreatment

Thehazhnan K. Ponnaiyan, Chemical & Environmental Engineering, University of Toledo, Toledo, OH

Ananth P. Dadi, SuGanit Systems, Inc., Toledo, OH

Fei Zhao, Chemistry Department, University of Toledo, Toledo, OH

Christa M. Graham, Chemistry Department, University of Toledo, Toledo, OH

Constance Schall, Chemical & Environmental Engineering, University of Toledo, Toledo, OH

Jared Anderson, Chemistry Department, University of Toledo, Toledo, OH

Sasidhar Varanasi, Chemical & Environmental Engineering, University of Toledo, Toledo, OH

Many technical challenges exist in the development of commercially viable biorefining processes for plant (lignocellulosic) biomass, although it is an abundant, inexpensive and sustainable source of organic carbon for the production of fuels and chemicals. Pretreatment of biomass and enzymatic hydrolysis are critical steps, and most of the current pretreatment methods fail to effectively disrupt the crystallinity of cellulose, which is a major barrier for efficient enzymatic hydrolysis. Ionic liquids (ILs) being non volatile, non-flammable, recyclable and designer friendly, are gaining wide recognition as green solvents, and their unique solvating properties make them ideal for pretreating lignocellulosic substrates. IL pretreated lignocellulosic substrates like corn stover (softwood) and poplar (hardwood) are shown to produce enhanced enzymatic hydrolysis yields of glucan to glucose and xylan to xylose. However, economic viability of this promising pretreatment technique materializes only when the process solvent (IL) is completely recovered and recycled. After incubation with biomass, the IL is separated from the latter via dissolution in a suitable anti-solvent (water/ethanol). Subsequently, IL can be recovered from the anti-solvent via evaporative and/or membrane based separations. It is of importance to quantify the residual anti-solvents in the recovered IL, to establish their permissible levels.

This study reports: (1) the use of ILs as non-derivitizing green solvents for the pretreatment of lignocellulosic biomass for subsequent bio-ethanol production, (2) the recovery of ILs using an evaporative separation scheme, along with a concomitant assessment of the efficiency of the recycled IL for biomass-pretreatment over several cycles of recovery and reuse, and (3) the establishment of acceptable levels of anti-solvent in the recycled IL (without compromising on its effectiveness as a pretreatment agent), and the development of analytical methods to accurately measure these concentrations.

See more of this Session: Integrated Biorefineries for Lignocellulosic Biomass
See more of this Group/Topical: Forest and Plant Bioproducts Division

Acetic Acid Removal from Pre-Pulping Wood Extract

Aymn Abdulrahman, Byung-Hwan Um, Audrey Polifka
Adriaan van Heiningen, G. Peter van Walsum,

University of Maine, Department of Chemical and Biological Engineering
Orono, Maine 04469 USA

Abstract

Pre-pulping extraction is a means of deriving a hemicellulose-rich process stream from the front end of a Kraft pulp mill. When the extraction is carried out using green liquor, pulp quality and quantity can be retained while still releasing hemicelluloses and acetic acid for recovery as bioprocessing feedstock or chemical products. The acetic acid that is present in the wood extraction is inhibitory to microorganisms and can inhibit fermentation. It is also a commodity chemical that may provide retain sufficient value to justify recovery and purification.

In this study, a liquid-liquid extraction method is applied to extract acetic acid from a green liquor pre-pulping hardwood extract. The acetic acid removal process takes place as an initial step prior to the fermentation process. An organic solution, such as Trioctylphosphine oxide (TOPO) in Un-decane, is prepared and mixed with wood extract. Next the extract phase is centrifuged in order to separate the aqueous and organic phases. The aqueous phase is sent on to fermentation while the organic phase is sent on to distillation to separate acetic acid from the organic phase. Finally, the organic solvents are recycled back to the extraction. Results present the extraction and recovery efficiencies. Preliminary comparisons are made with other potential separation technologies.

Introduction

Fossil fuels, generated from crude oil, coal and natural gas, have been reported to be one of main causes of the global warming, greenhouse gas emissions and associated environmental problems. Moreover, increasing crude oil prices and petrochemical fuel demands, and decreasing fuel supplies and foreign oil imports lead researchers to search for renewable, reliable and alternative energy sources. As a renewable, widespread, cheap and sustainable fuel, biomass-derived fuels such as ethanol are more attractive because of their potentially plentiful supply and relatively low cost. Biomass conversion is also utilized to produce valuable and important by-product chemicals such as acetic acid [1,2,3].

Biomass Conversion

Biomass has been used for a long time as a primary source of energy all over world. Today approximately 12% of the world's energy supplies are produced from biomass and in many developing countries the percentages are reaching to 40% and 50%. Biomass has been used or converted to produce renewable fuels and energy sources in solid, liquid and gas phases.

The United States is one of many countries trying to use biomass conversion processes to produce bio-fuels and energy in an effort to reduce fossil energy consumption. As an example of that, it has been proposed by the U.S. department of Energy [5] that about 512 million dry tons of biomass residues in the United States could be converted to liquid fuels and 10% to 50%

of transportation fuels could be replaced by biomass fuels between 2010 and 2050.

Lignocelluloses Composition

Wood lignocelluloses mainly consist of a mixture of polymerized carbohydrates (cellulose and hemicelluloses), lignin, extractive components and ash. Wood can be generally categorized as hardwood or softwood and the physical and chemical properties between them vary [4].

Hemicelluloses are linear and Y branched polymers that are shorter than cellulose with a degree of polymerization of 50-200 units. They consist of three hexoses (D-glucose, D-galactose, and D-mannose) and two pentoses (D-xylose and L-arabinose). Hemicelluloses work as a linkage between cellulose and lignin [3]. There are two common types of hemicelluloses in wood: Xylans and galactoglucomannans, (Figures 1 and 2). These two are varying in terms of compositions and properties from one species to another.

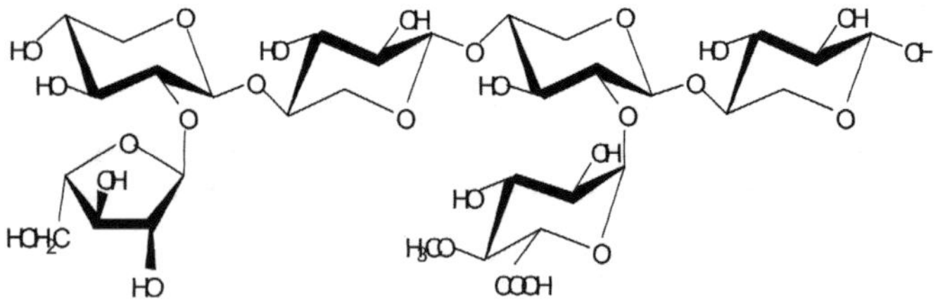

Figure 1: Structural representation of softwood xylan. Containing acetyl groups in 4-O-methyl α-D glucopyranosyluronic acid (bottom right) [9]

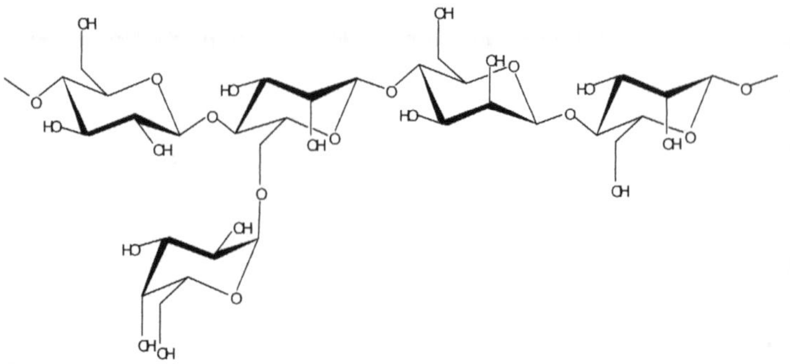

Figure 2: Structural representation of softwood galactoglucomannans [9]

Kraft Pulping Process

As an important process for pulp and paper technology, the Kraft pulping process is used widely all over the world and especially in the state of Maine. For more than 100 years, it has to be the most economical chemical pulping process. In 1998, about 3.8 million tons of products were produced from pulp manufacturing [4]. In the Kraft process approximately 50% of the wood weight is dissolved into pulping liquor. This solution is primarily composed of hemicelluloses and lignin [7].

About half of wood hemicelluloses can be removed in the Kraft pulping process because hemicelluloses have a lower degree of polymerization (50-200) that can be more easily dissolved into the aqueous solution than other lignocellulosic components [7]. Hemicelluloses are usually burned to produce energy during the chemical recovery process [7], but because they have a heating value of about 13.6MJ/kg, which is only about half the heating value of lignin [9], they may find a more optimal use as a feedstock for chemical conversion.

Pre-pulping extraction process

The primary processes to produce ethanol from extracted hemicellulose are shown in Figure 3. These processes are: hemicelluloses extraction, hydrolysis to release sugars using sulfuric acid, filtration to extract lignin, liquid-liquid extraction to recover acetic acid and furfural, gypsum separation and fermentation of C5-C6 sugars to produce ethanol [6].

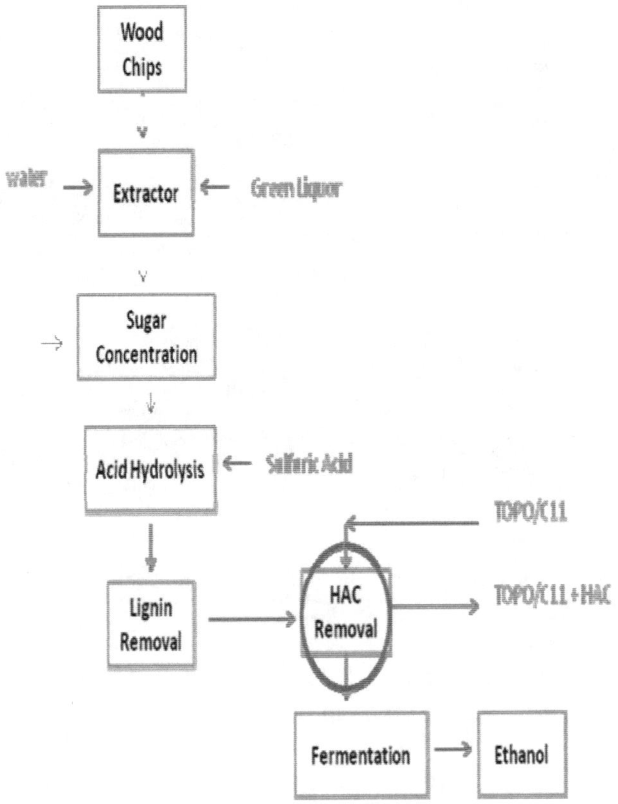

Figure 3: Overall processes to produce ethanol

There are three common processes to convert lignocelluloses to ethanol: hydrolysis, either by acid or by enzymes, and fermentation. The hydrolysis process is used to break the crystalline structure of the lignocellulosic material (cellulose and hemicelluloses) and release monomer sugars such as glucose and xylose in order to make them more digestible in the fermentation process. Hydrolysis can be accomplished using acids, such as sulfuric acid, or cellulolytic enzymes. These sugars are then fermented using microbes such as Baker's yeast to produce ethanol [3]. Degradation products from the hydrolysis process, such as furfural, hydroxymethyl furfural (HMF) and acetic acid could be harmful for some fermentation organisms. Xylose is converted to furfural while glucose is converted to HMF.

Fermentation affected by acetic acid

Acetic acid, CH_3COOH, is an important biomass product which has a high marketing selling price (approximately double the ethanol price) [6]. It is released when the acetyl groups of hemicelluloses, especially hardwood xylans, are hydrolyzed. Acetic acid can be used in many industrial applications such as chemicals, light industry, textiles, pharmaceuticals, printing/dyeing, rubber, pesticides, photographic chemicals, electronics, and food processing (as vinegar) [1].

Acetic acid is a typical weak acid and is partially ionized in its solution according the following reaction.

$$CH_3COOH \rightleftarrows H^+ + CH_3COO^-$$

The thermodynamic equilibrium constant, K_a can be defined by:

$$K_a = \frac{[H^+][CH_3COO^-]}{[CH_3COOH]}$$

Acetic acid is one of the components that could inhibit fermentation organisms, which affects ethanol production efficiency. Acetic acid can penetrate the bacteria cell walls and acidify the cytoplasm. As a result of that, the protein gradient across the cell membrane is disrupted [10]. Therefore, it is recommended to extract acetic acid early before reaching the fermentation stage. Liquid- liquid extraction and distillation are two common processes to extract acetic acid from hemicellulose extracts.

Method

In the University of Maine's pilot plant, hemicellulose extracts are prepared by loading 2 kilograms of commercial freshly mixed northeastern hardwood chips sized between 16 and 22.6 mm, green liquor (contain 1 - 4% of equivalent Na_2O) and water (including wood moisture) at overall ratio of 4:1 liquor/dry weight into a 20 liters rocking digester. The system is then agitated at a speed of 2 rpm for 110 minutes and heated at 160 °C. Next, the system is cooled below 100 °C and the freely drained liquor is collected. The composition of the green liquor and hot water extract are shown in Table 1 [7].

Table 1: Composition of 3 % Green Liquor (GL) and Hot Water (HW) Extracted Northern Hardwood

Composition of Extract	GL Extract[2] [g/L]	GL Extract[3] free monomers [g/L]	HW Extract[2] [g/L]	HW Extract[3] free monomers [g/L]
Glucose	0.36	0.09	1.01	0.09
XMG[1]	1.62	0.11	9.06	0.65
Arabinose	0.24	0.02	1.37	1.07
Acetic Acid	9.40	8.32	3.47	1.27
Total Sugars	2.22	0.22	11.44	1.81

[1] xylose, mannose, and galactose, [2] secondary hydrolyzed with 4 % sulfuric acid, [3] analyzed without secondary hydrolysis.

Materials

The chemicals used were: Trioctylphosphine oxide (TOPO) ($[CH_3(CH_2)_7]_3PO$, Mol W. 386.63, 99% purity, Sigma) , Un-decane ($CH_3(CH_2)_9CH_3$, Mol W. 156.3, 99% purity, Sigma), Trioctylamine (TOA) ($[CH_3(CH_2)_7]_3N$, Mol W. 353.67, 98% purity, Sigma) , Octanol ($CH_3(CH_2)_7OH$, Mol W. 130.23, 99% purity, Sigma), and acetic acid (CH_3CO_2H, Mol W. 60.05, 99.99+% purity, Sigma).

Liquid-Liquid Extraction

The liquid-liquid extraction method is applied first to clean aqueous solution (acetic acid and water) and then to real extract (green and hot water) to extract acetic acid.

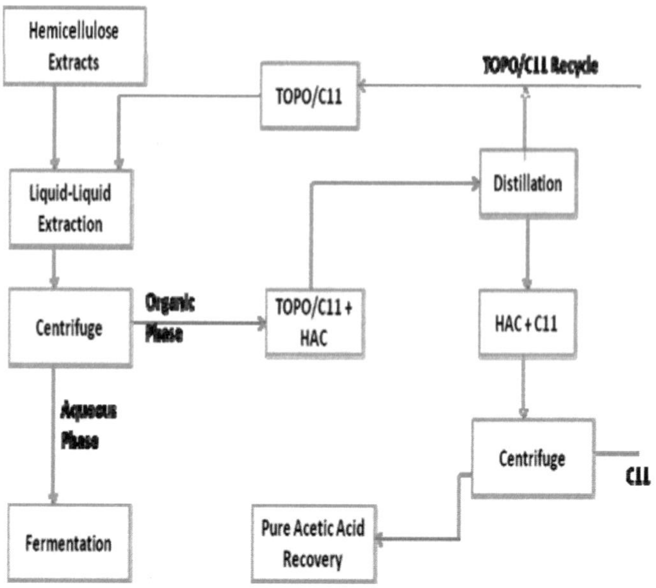

Figure 4: Liquid-liquid extraction for acetic acid followed by distillation system

TOPO has a high molecular weight and works as an extraction agent to remove fatty acids like acetic acid from aqueous solutions. TOPO has strong hydrogen bonding properties that induce the carboxylic acid to transfer into the organic phase. In addition, TOPO is characterized by its excellent stability, high boiling point and low solubility in water [11].

An organic solution of TOPO is prepared by mixing TOPO with undecane ($C_{11}H_{24}$). Then a typical extraction procedure is as follows: 10.0 mL of the aqueous phase (green liquor- or hot water-generated hemicelluloses extract) which contain a certain amount of acetic acid was shaken vigorously with 10.0 mL of the organic phase (undecane as a diluent containing a certain amount of TOPO) for 36 minutes at various temperatures. Samples of both phases were analyzed immediately after the phase separation. The concentration of acetic acid in the organic phase was determined by titration using 0.05 M NaOH with phenolphthalein as indicator while the concentration of acetic acid in the aqueous phase was determined by High Performance Liquid Chromatography (HPLC). Gas Chromatography (GC) was used to detect quantities of organics in the organic phase.

Distillation

The organic phase after extraction process is taken to a distillation system to separate acetic acid from the organic phase. The TOPO in the organic phase after distillation is recycled again to reduce the cost.

Results

The following tests have been done for this system using both solvents TOPO and TOA:

- Effect of varied concentrations of acetic acid in aqueous solution
- Effect of varied concentrations of solvent in organic solution
- Effect of varied PH
- Effect of varied ratios of organic to aqueous phase
- Recycle efficiency of solvent

This extraction is used for a clean solution (acetic acid and water) with varied of acetic acid concentrations from 10 to 100 g/L (figure 5). The organic to aqueous ratio was 1:1 and PH was 2 at reaction temperature was 70 C°.

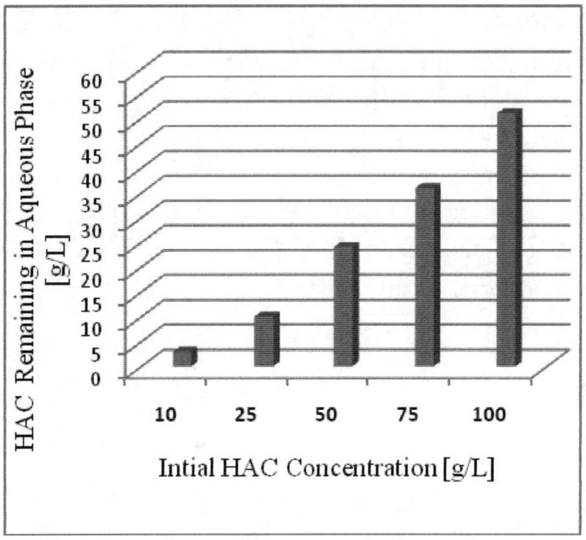

Figure 5: Effect of varied concentrations of acetic acid in aqueous solution

The extraction efficiency is shown to decrease as the initial acetic acid concentration increases. This is because the organic solution reaches a saturated condition with no more capacity to bind with acetic acid.

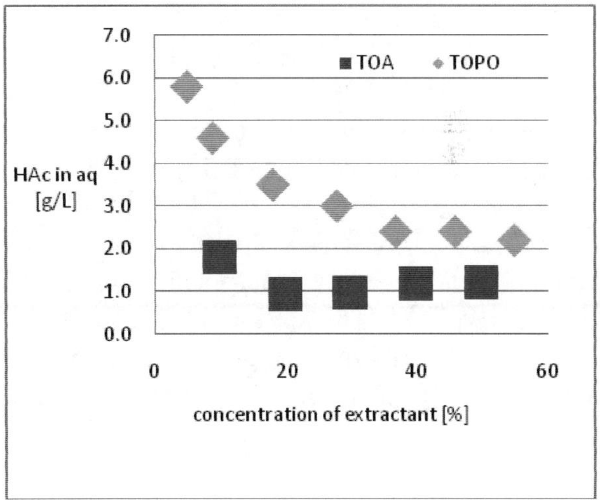

Figure 6: Effect of varied concentrations of solvent in organic solution

Figure shows the effect of using different concentrations of TOPO and TOA. The system is tested for a simple water and acetic acid system. Both the solvents are varied from 10 to 50 g/L in their diluents.

TOPO is shows good extraction when it has a concentrationof 37% (w/v) of or more in Un-decane, while TOA shows high efficiency down to 20%in its solvent. In addition, the maximum extraction percentage of TOA is 90% while TOPO is 70%. Therefore, TOA superior to TOPO at extracting acetic acid .

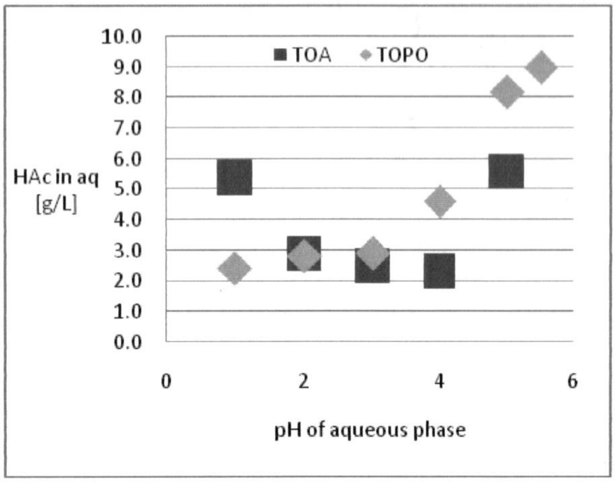

Figure 7: Effect of varied pH

Figure 7 is shows a good extraction when the pH is low. Therefore, TOPO has optimal extraction when pH is at or below 3 and shows less efficient extraction when pH is above 3. On other hand, TOA is shows good extraction up to a pH of 4 and only deteriorates above pH 4. Since the green liquor extract can have an approximate pH of 4, using TOA solvent may eliminate the need to lower pH to get a good extraction while using TOPO does require reduction in pH..

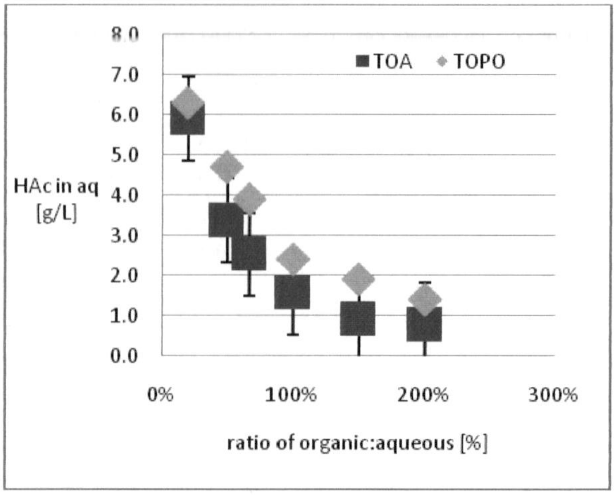

Figure 8: Effect of varied ratios of organic to aqueous phase

The extraction efficiency is affected with organic to aqueous ratio. When the ratio increases, the extraction efficiency also increases. However extraction efficiency appears to plateau near a 100%.ratio, Therefore, the best ratio is found to be 1:1.

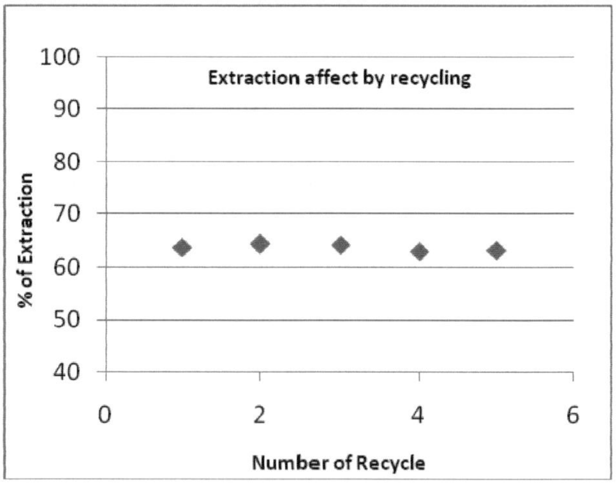

Figure 9: Recycle efficiency of solvent

This test has been done for TOPO solvent is very expensive and therefore requires good recovery. Figure 9 illustrates the extraction performance through 5 recycles of the solvent through distillation. This demonstrates that on a clean system (water and acetic acid) TOPO in undecane appears to be readily recoverable. At the time of this writing, the recyclability has not yet tested on real extract such as green liquor extract.

Table 4: Comparisons of using TOPO and TOA for Liquid-Liquid extraction

	TOPO	TOA
Max Extraction	70 %	90 %
Acetic acid concentration in aqueous	10 g/L	10 g/L
Solvent concentration in organic	37 %	20 %
PH	≤ 3	≤ 4
Ratio	1:1	1:1
Reaction Temperature	70 C^0	70 C^0
Reaction Time	1 hr	1 hr
Recyclability	Stable	-

Note: The extraction parameters are used for a clean system

Future Works

- Test the extraction and recyclability of TOA on model solutions and wood extracts.
- Test the recyclability of TOPO solvent for real extract system

Conclusion

Liquid-Liquid extraction is used to extract acetic acid from aqueous solution. The extraction was done first for a clean system (acetic acid in water) and then for wood extracts (green liquor extract). The recyclability of TOPO solvent was done for the clean system to reduce the cost of the solvent.

TOPO shows 70% extraction for clean system and 64% for wood extract system. While TOA is shows 90% acetic acid removal in the clean system. The best extraction parameters for this system are :

- The concentration of acetic acid in aqueous solution is to be 10 g/L.
- The concentration of TOPO in organic solution is to be 37%, while TOA is to be 20%.
- PH is to be ≤ 3 for using TOPO and ≤ 4 for TOA
- The ratio of organic to aqueous is to be 1:1

Finally recyclability of TOPO for the clean system is showing stable performance with no affect to extraction efficiency.

References

[1] X. Xu, Direct Conversion of Carboxylate Salts to Carboxylic Acids Via Reactive Extraction, M.S. Thesis, Texas A&M University, (2008).

[2] A. van Heiningen, Converting a Kraft Pulp Mill into an Integrated Forest Biorefinery, PAPTAC. 107 (2006) 141-146.

[3] F. Agbogbo, Anaerobic Fermentation of Rice Straw and Chicken Manure to Carboxylic Acids, PhD Thesis, Texas A&M University, (2005).

[4] M.F Demirbas, Current Technologies for Biomass Conversion into Chemicals and Fuels, Energy Sources Part A. 28 (2006) 1181-1188.

[5] B. Han, W. Carvalho, L. Canilha, S.S. da Silva, J.B.A. e Silva, J.D. McMillan, S.R Wickramasinghe, Adsorptive Membranes vs Resins for Acetic Acid Removal from Biomass Hydrolysates, Desalination. 193 (2006) 361–366.

[6] H. Mao, J.M. Genco, A. van Heiningen, H. Pendse, Technical Economic Evaluation of a Hardwood Biorefinery Using the "Near-Neutral" Hemicellulose Pre-extraction Process, J. Biobased Mater. Bioenergy. 2 (2008) 1-9.

[7] S.L. Walton, Biological Conversion of Hemicelluloses Extracted from Hardwood,

Unpublished Results, University of Maine, (2008).
[8] B.J.W. Cole, Wood Chemistry, Class Notes, University of Maine, (2009).
[9] E. Sjöström, Wood Chemistry: Fundamentals and Applications, second ed., Academic Press, California,1993, pp.65.
[10] C.M. Takahashi , D.F. Takahashi, M.L.C. Carvalhal, F. Alterthum, Effects of Acetate on the Growth and Fermentation Performance of *Escherichia coli* KO11, Appl. Biochem. Microbiol. 81 (1999) 193-203.
[11] H.F. Al-Mudhaf , M.F. Hegazi, A.I Abu-Shady, Partition Data of Acetic Acid Between Aqueous NaCl Solutions and Trioctylphosphine Oxide in Cyclohexane Diluent. *Sep. Pur. Technol.* 27 (2002) 41–50.

Key Metric Comparison of Five Cellulosic Biofuel Pathways

by B. A. Thorp [(1)]

Purpose

Cellulosic biofuels are relatively new products except for a handful of sulfite pulp mills that have fermented their red liquor to obtain ethanol as early as 1909.[(2)] Sulfite pulping is obsolete and the modern pathways for cellulose are pyrolysis, gasification, acid hydrolysis and enzymatic hydrolysis. While the pathway from pyrolysis to transportation fuel is not complete, the pathway to fuel for industrial boilers will be included for reference. Commercial efforts in cellulosic biofuels are new and data to compare a specific offering to others within a class or to compare one pathway to another do not exist. Data are lacking for a complete comparison. This paper will start building a comparison metrics framework for cellulosic biofuel pathways. Others will need to further build on and update this comparison. The metrics for ethanol from corn and biodiesel from transesterification are well known and can be added as needed. Later, advanced biofuels like those derived from barley, sorghum , algae, sugar cane and hybrid pathways can be added.

The metrics chosen are capital in dollars per annual gallon, yield in gallons per ton, the co-production of MAJOR secondary products, operating cost per gallon, and the selling price of the biofuel as a percentage of gasoline. These key metrics focus on conversion to liquid transportation fuels. Conversion to chemical products can improve overall economics. The desired yield is BTUs sold divided by BTUs purchased but there is not yet enough published to determine this key variable. It is shocking to find so little data or understanding of this metric as biorefineries are fundamentally a BTU transforming facility. All petrochemical refineries know, track and optimize this metric. The operating cost per gallon is highly dependent of feedstock cost and that is a function of time, location and type. By establishing a selling price as a percentage of gasoline the time variation of commodity oil pricing is eliminated.

There are a large number of technology approaches. Grouping of technologies will be required to simplify metric development shown in Figure 1 below. Gasification can be used with two very different Gas-to-Liquid (GTL) pathways so each will be discussed.

Figure 1

Metrics of Conversion Pathways

*data not available for the desired BTU yield

Conversion Pathway	Capital Cost ($ per ag)	Yield* (gal per d ton)	Secondary Products	Op. Cost ($ per gal)	Selling Price (% gasoline)
Pyrolysis					
Gasify then convert to Ethanol					
Gasify then F-T to Diesel					
Acid Hyd. then Fer.					
Enz. Hyd. then Fer.					

Introduction

Most qualified forecasters conclude that global energy prices will increase as demand stresses supply. The only difference in the forecasts are time and intensity of the increase. Some forms of energy like electricity have alternate production sources. Those sources include wind, solar, hydro and nuclear. The only non-petrochemical alternatives for liquid transportation fuel are biofuels or switching engines from internal combustion to electric power. Hydrogen may appear as a fuel in the future but most predict that it is decades away. Now, biofuels become one of the factors which are critical to the economic health and even survival any oil importing nations so that biofuels can supplement oil for transportation needs and industrial process needs.

Understand the relationship of raw materials, markets and technology is critical. For any given location the quantity of raw material that exists or can be grown within an economical harvest radius is fixed. This makes raw material an independent variable which must be quantitavely understood. Location can also have a major impact on the type of biofuel that can be effectively marketed. For example, some regions have an excess of gasoline and ethanol while others have shortages or higher logistics costs. Considerations such as these make markets an independent variable which must be quantitavely understood. If it is possible to define the raw material and the market, then conversion technology becomes a dependent variable and should be selected because of its suitability to convert the selected raw material to the biofuel selected for a market.

After identifying raw materials and markets, the next step is identifying the most promising pathway, then identifying the promising technology within a pathway. The last step is to validate all choices. This means that there will not be a single technology of choice but several technologies of choice for combinations of raw materials and markets.

Figure 2 below is an illustration of the optimum area among raw materials, markets and technologies.

Figure 2

Optimum Area

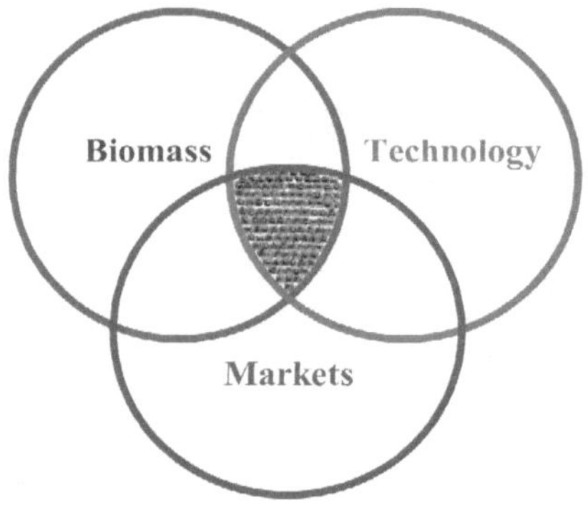

Pyrolysis

The first pathway is pyrolysis of biomass. Producing bio-oil is an old technology in which biomass is volatized by high temperatures, the condensed into a liquid. Pyrolysis oil is acidic having a pH in the range of 2 to 3, has a shelf life as it continues to slowly react and has a lower BTU content because it contains low BTU components like water. The process is outlined in Figure 3 below.

Dynamotive, UOP/Honeywell and others are developing new and improved processes to convert bio-oil into transportation fuels. While progress has been made and reported, these developments are not far enough along to establish key metrics. However, it is important to know the key metrics of the base process.

Figure 3

Pyrolysis of Biomass to Bio-oil

- Green biomass is harvested and sized. For sawdust , no preparation is needed.
- Rapid thermal pyrolysis volatizes the mass.
- Char is rejected with a simple cyclone.
- The volatiles are quenched/condensed and collected.
- Products from sawdust are ~70% Bio-oil (hi yield), ~15% Char and ~15% non-condensable gasses.
- This is an old process that has been used in laboratories for decades.

There has been ample commercial activity using the pyrolysis technology. A summary of commercial Plants in North America include those shown in Figure 4.

Figure 4

Pyrolysis Plants Include

- There are several pilot plants in Canada including Guelph and West Lorne, Ontario.
- The Ensyn technology is used in 4 Red Arrow plants where flavors are extracted from the Bio-oil which then becomes a by-product.
- The first commercial plant announced *2009/10* in Willow Springs, MO by Dynamotive. It has a cost of $24 million and will produce 12 million gpy Bio-oil sold for commercial boilers.
- Overseas plants have also been announced.

Since pyrolysis is an older technology a number of the metrics are documented. Advantages include that is a known process with demonstrated costs and produces a biofuel with known properties. The major advantages and disadvantages include those shown in Figure 5 [(3)].

Figure 5

Pyrolysis Major Strengths and Weaknesses

- **Major strengths include:**
  Low capital: ~$2.00 per annual gallon
  Any tree or crop species
  Simple preparation of raw material
  Hi yield: ~170 gpt, and low operating cost
- **Major weaknesses include:**
  Unstable as reactions slowly continue; shelf life issues
  Low BTU content as it contains ~15% water
  Very acidic, cannot be used in all boilers
  Therefore, lower selling price: ~50% of gasoline

Gasify Biomass & Convert Syngas to Ethanol

Gasification, or heating biomass with controlled oxygen, is well known process. There are numerous commercial industrial processes in North America.[(4)] The resulting syngas must be cleaned to specifications determined by the downstream process which will convert the syngas to liquid. After cleaning the syngas will consist of C, CO_2, H_2 and only traces of other compounds. The syngas can be converted to ethanol chemically via Mixed Alcohol Synthesis (MAS) (Range Fuels), packed bed catalytic reactors (Fulcrum), or fermentation (INEOS and Coskata). In addition to gas cleanup, the GTL processes can require compression to achieve the needed temperatures and pressures. Fermentation does not require compression but does require an "incubation tank farm" sized with enough units to match gas production capacity. A generalized process description is shown in Figure 6. Diverse raw materials can be gasified and include forest slash, spent crops, energy crops, and municipal solid waste (MSW) and sorted MSW (SMSW). Regulation prohibits sorting inside the landfill. Diverse feedstocks can impose additional requirements for storage, preparation and drying prior to gasification.

Figure 6

Biomass Gasification Followed by Conversion to Ethanol

- Biomass is harvested, sized and dried as required by gasification technology. MSW is sorted and prepared outside the landfill as required by regulation. Sorted MSW goes to plant at no/low cost.
- Feedstock is then gasified and the syngas cleaned. For SMSW some prefer downdraft partial oxidation followed by plasma arc. Heat can be recovered as syngas is cooled and cleaned.
- The syngas is converted to ethanol by MAS or proprietary packed bed catalytic reactors. Alternately it can be fermented. Heat can be recovered from some processes. Fermentation requires little/no additional heat.
- Distillation/dehydration is required for most processes. Membrane separation has lower costs but requires demonstration.

Recent Department of Energy (DOE) grant solicitations have augmented commercial activity. There are several pilot plants and at least six announced and partly-funded commercial plants which are shown in Figures 7 and **7a.**

Figure 7

Biomass Gasification Plants Include

- Range Fuels is constructing a phase 1 plant in Soperton, GA which is scheduled to make 8 million gallons ethanol and 2 million gallons methanol.
- The first SMSW plant may be Sierra BioFuels (Fulcrum Energy), McCarran NV, *scheduled to start in 2011.* This plant will produce 10.5 million gpy ethanol + some solid waste for landfill. Permits have been obtained and engineering complete. Cost estimated to be ~$120 million.
- Enerkem has announced financing to gasify SMWS and produce 9.5 million gpy and 20 million gpy respectively in Edmonton, Canada and Pontotoc, MS.
- Coskata is building a demo plant in Warren County, PA which will ferment syngas to produce 40,000 gpy ethanol. They are actively looking for a commercial site.

Figure 7a

Biomass Gasification Plants Include
(continued)

- Powers Energy has announced a 32 million gpy plant in Indiana converting MSW to ethanol using INEOS technology.
- Gulf Coast Energy has a 20,000 gpy demonstration plant in Livingston, AL based on Pearson technology funded for years by DOE. Higher yields reported in electronic media have not been validated.

The major strengths and weaknesses that have been identified to date from proposed installations of gasifying biomass for the production of ethanol include those shown in Figure 8.

Figure 8

Biomass Gasification then Ethanol Major Strengths & Weaknesses

- **Major strengths include:**
  - Feedstock diversity
  - SMSW feedstock operating cost can be as low as \$0.24/gal
  - Lower capital cost claimed by those using fermentation
  - Yield (80 gpt to claims of 111 gpt)
  - Smaller units can be self-financed
- **Major weaknesses include:**
  - Some approaches lack extended pilot line runs
  - Pilot lines are propriety, not always available
  - Sierra BioFuels will be a pilot/demo site
  - Plasma arc gasifiers are difficult to operate
  - Some syngas conversion technology unproven
  - Product selling price ~70% of wholesale gasoline (BTU basis)
  - Many plants are stand alone and appear under optimized for heat recovery

Gasification followed by Fischer-Tropsch (F-T)conversion

This process was first used commercially by the German military in WWII to produce ~5 million gallons per day when oil supplies were curtailed. The raw material was coal. The next commercial use was in the 1960's by South African Synthetic Oil Ltd (SASOL) as a way for South Africa to survive under the embargo. The Secunda facility still produces ~6,720,000 gallons per day. The US challenge is to replace coal with biomass and use a gas cleanup process which is optimum for biomass. Figure 9 shows the general process.

Figure 9

Biomass Gasification then Fischer-Tropsch Conversion to Diesel & Wax

- Size and dry biomass
- Gasify (Choren, GTI, TRI & Rentech)
- Gas cool down, heat recovery and cleanup
- Compress and pass through F-T
- Burn naphtha and F-T tail gas
- Generate green power for facility or grid, recover steam for a user-host
- Separate diesel and wax. Wax can also be converted to diesel

There are several plants being designed for the U.S. and the major ones are shown in Figures 10 and 10a. Choren is also shown because their plant in Germany is being commissioned and is an important benchmark.

Figure 10

Biomass then F-T Plants Include

- Choren has a 4.2 million gpy stand alone plant in Freiberg Germany with a yield of ~60 gpt. Thermal Eff. ~41%.
- NewPage won a $30 million DOE grant to build an integrated 5.5 million gpy D&W facility in Wisconsin Rapids WI. The mill also gets 200 million BTU/hr stream, 123 million BTU/hr hot water and 14 million BTU/hr gas for the lime kiln. Thermal Eff. ~70%. Cost ~$90 million. *Start date not announced.*
- Flambeau River Biofuels won a $30 million DOE grant to build an integrated 17 million GPY D&W facility in Park Falls WI. The mill also gets 5.8MW power and 190 million BTU/hr steam. Thermal Eff. ~60%. Cost $260 million. *Start date not announced, but 2012 likely.*

Figure 10a includes configurations which have not been discussed so far. That is some mixture of biomass and fossil fuel which is available in the area. The reason this will be important in the future is that biomass availability may limit biomass biorefineries to a size of ~50 to 100 million gpy. By incorporating 50% fossil fuel the size could be increased to 100 to 200 million gpy, thereby gaining economy of scale. Capital intensity is decreased 34% by increasing plant size from 25 to 100 million gpy and 20% by increasing capacity from 50 to 100 million gpy.[5] Some of these configurations could possibly meet the Green House Gas (GHG) standards in the 2007 Energy Independence and Security Act (EISA). This approach has technical, environmental and social challenges.

Figure 10a

Biomass then F-T Plants Include

(continued)

- Rentech will build a 1.5 million gpy plant in California and has a signed contract with airlines to use the fuel in ground operations.
- Rentech proposed a 95% petroleum coke and 5% biomass for Phase 1, 24 million gpy plant in Natchez, MS by 2011. Phase 2 is 412 million gpy.
- Rentech - Mingo proposed to make 441 million gpy in Mingo County, WV from coal and wood waste.
- Rentech - Solena proposed to make 220 million gpy jet fuel from BOIMASS in northern CA using the Solena gasifier.

The strength and weaknesses are well known due to commercial experience and recent biomass design. These systems can have greatly improved economics when the heat from syngas cool down/clean-up and from F-T tubes can be recovered and sold.

Figure 11

Biomass Then F-T
Major Strengths & Weaknesses

- **Major strengths include:**
  Process commercial in "components"
  Uses the whole plant/tree
  Can use mixed biomass
  Can sell waste heat as well as diesel & wax
  Sulfur free diesel can sell for ~150% gasoline
  Pilot and demo plants proprietary but available

- **Major weaknesses include:**
  High capital cost
  Lower yield but high thermal eff., if integrated
  Has high power requirement for F-T unit

Strong Acid Hydrolysis

Like pyrolysis, this is an old technology. In this pathway there can be mild acid hydrolysis first stage treatment if the raw material has hemicellulose in quantity. The remaining biomass is subjected to concentrated sulfuric acid to convert the cellulose to glucose which is then fermented to ethanol or chemical products. Sulfuric acid is recovered, purified and reused. Lignin will be a residue and can be collected and sold. It can also be burned to generate process heat if the capital for the boiler can be justified by the savings. This has not been part of the projects listed. Figure 12 shows a generic outline of the process.

Figure 12

Strong Acid Hydrolysis

- Can have "mild acid" pretreatment to remove hemicelluloses if raw material has them in quantity
- Reclaims concentrated sulfuric acid, applies it to the raw material converting cellulose to glucose, then uses chromatographic separation to isolate the sugars.
- The sugars are then fermented to ethanol with conventional technology, including distillation and dehydration.
- If the raw material contains lignin there is discussion about using it to produce energy. Otherwise natural gas is used.
- Yield is low in the 60 to 70 gpt range.

The most modern facility is the demonstration plant outlined in Figure 13 below. This is followed by a very old plant in Russia and a modern plant targeted for California.

Figure 13

Acid Hydrolysis Plants Include

- There is a 9,000 gpy demonstration plant in Izumi Japan (southern tip). Various yields reported but 64 gpt seems reasonable. SEKAB has a pilot plant in Ornskoldsvik, Sweden.
- Tavada, Siberia has produced 3.3 million gpy of ethanol from 70% wood, 20% wheat straw and 10% beet molasses since 1943. The 1997 cost estimate was $5.67 per gallon with raw materials and energy (coal) totaling 70% of costs.
- In 2007 BlueFire Ethanol won a $40 million dollar DOE award to build a 100 million dollar plant using sorted MSW. Yield reported to be 68 gpt. BFE has applied for a DOE Loan Guarantee to build a 3.2 million gpy plant in California.
- There is discussion of a plant in South Korea.

Since this technology has been studied for decades the strengths and weaknesses are well understood and shown in Figure 14. A major strength is the ability to use Sorted Municipal Solid Waste and perhaps get a tipping fee from urban areas where landfill capacity is critical. This is an example of locating the raw material then selecting the most appropriate technology for conversion.

Figure 14

Acid Hydrolysis
Major Strengths & Weaknesses

- **Major strengths include:**
  Known process, over 100 years old
  Excellent pilot/demo plant in Japan
  Can use low cost raw material like SMSW

- **Major weaknesses include:**
  Capital cost high due to use of acid
  Operating cost high
  Yield on the low side of "average"

Enzymatic Hydrolysis followed by Fermentation to Ethanol or Other Chemicals

This is a newer technology with a great deal of emphasis, research and federal funding over the last decade or so. So far, lignin has not responded to enzymatic hydrolysis so it becomes a by-product. Lignin can also slow the enzymatic process so many companies have focused on low-lignin-content biomass like corn cobs, corn stover, spent sugar cane and various straws. This minimizes issues with lignin. Figure 15 below outlines a generalized process.

Figure 15

Enzymatic Hydrolysis

- Adequate feedstock storage for ~10 months (for crops)
- Feedstock preparation
- Mild acid hydrolysis for separation/removal of hemicellulose
- Fermentation of 5 carbon sugar to ethanol
- Enzymatic hydrolysis of cellulose to glucose
- Fermentation of glucose to ethanol
- Removal of lignin for sale or energy use
- Distillation
- Dehydration

Consistent with the emphasis there are a growing number of pilot, demonstration facilities and proposed commercial facilities. The major ones are shown in Figures 16 and 16a. It is likely that there will be additional announcements, sometimes on a weekly basis.

Figure 16

Enzymatic Hydrolysis Plants Include

- Verenium, Jennings, LA demo plant converting spent sugar cane and switchgrass to 1.4 million gpy ethanol. Also has pilot plant.
- KL Energy has a 1.5 million gpy demonstration plant in Upton, WY since 2008 using pine chips
- Mascoma recently started a 200,000 gpy pilot plant in Rome, NY.
- Verenium announced a ~$275 million commercial plant sized for 36 million gpy for Highlands County, FL. Partners include Lykes Bros. and BP Oil. Received $7 million grant from Florida.

Figure 16a

Enzymatic Hydrolysis Plants Include
(continued)

- Mascoma received a $30 million DOE grant and a $24 million Michigan grant to build a 40 million gallon per year woody biomass to ethanol plant in Michigan. Estimated cost was ~$300 million.
- Poet has a self funded 20,000 gpy pilot line in Scotland, SD using a proprietary process to convert corn cobs to ethanol. The 30 million GPY commercial facility is planned for Emmetsburg, IA for which Poet has a DOE grant.
- Abengoa constructed a 20,000 gpy pilot line in York, NE using enzymatic hydrolysis of corn stover. Salamanca, Spain has a 1.3 million GPY demonstration plant. The 12 million gpy commercial plant is planned for Hugoton, KS for which they have a DOE grant.

Since this technology is relatively new and much is still proprietary the major strengths and weaknesses shown in Figure 17 below are more subjective.

Figure 17

Enzymatic Hydrolysis
Major Strengths & Weaknesses

- **Major strengths include:**
  Have operating demo plants with good scale to commercial
  Have strong partners
  Moderate capital cost

- **Major weakness include:**
  Must accumulate and store crop feedstock for long periods
  Moderate yield, reported at 80 gpt
  Enzyme cost typically high
  Selling price ~70% gasoline

There are also a number of hybrid technologies and two are highlighted in Figure 18 below. When demonstrated, Value Prior to Pulping (VPP) could add value to pulp mills because there is more value in converting the hemicellulose to ethanol or other chemicals than burning sugar in thermally inefficient recovery boilers.

Figure 18

Hybrid Technologies

- VPP-Value prior to HW pulping removes ~60% of hemicellulose before pulping, lowering pulping and bleaching cost. Hemicelluloses are then fermented to ethanol or other chemicals.
- Another hybrid combines enzymatic hydrolysis and gasification. The syngas derived from lignin is reacted with CO_2 from fermentation to produce more ethanol. The unique use of lignin increases yield.

The proposed plants are shown in Figure 19.

Figure 19

Hybrid Technologies Plants Include

- Old Town Fiber and Fuel, Old Town, ME won a 30 million dollar DOE grant to install the process at Old Town. Reports indicate a potential switch to a butanol process.
- Applied Biorefinery Sciences, Syracuse, NY is commercializing processes with higher extraction.
- ZeaChem, is constructing a 1.5 million gpy demonstration pilot plant in Boardman, OR.

The major strengths and weaknesses are shown in Figure 20. The VPP process can also be used on feedstock prior to pelletizing or combusting in a solid fuel boiler.

Figure 20

Hybrid Technologies
Major Strengths and Weaknesses

➢ Major strengths include:
- Low/no cost raw material for VPP
- VPP can be used on processes other than pulping
- Highest reported yield for ZeaChem

➢ Major weaknesses include:
- Both lack pilot/demo plants
- Selling price ~70% gasoline

Conclusions

There is sufficient publically available data and in the above presentation graphics to complete the Key Metrics boxes for Figure 21[(6)]. This figure is useful when studied for implications. Since a corn-ethanol facility can be completed for $2.00 to $2.50 per annual gallon, all pathways have relative high capital cost at this point in their development. At this time commercial success depends on capital reduction instruments like grants and/or unique synergies like tipping fees for raw materials or the sale of co-products like steam and power to host facilities.

Figure 21

Metrics of Conversion Pathways

*data not available for the desired BTU yield

Conversion Pathway	Capital Cost ($ per ag)	Yield* (gal per d ton)	Co-Products	Op. Cost ($ per gal)	Selling Price (% gasoline)
Pyrolysis	2	170	Char	~1	~50
Gasify then convert To Ethanol	4-6	80-100	TBD	1-2	70
Gasify then F-T to Diesel&Wax	6-8	50-70	Power & steam	1-2	~150
Acid Hyd. & Fer.	6-8	60-70	TBD	1.5-2.5	70
Enz. Hyd. & Fer.	4-7	70-80	TBD	1.4-2.5	70

Figure 22 shows general conclusions. Specific conclusions require a raw material availability analysis, a market analysis and a thermal balance for all facilities involved.

Figure 22

Conclusions

- Capital costs are the major barrier at current oil prices. Partnerships can be beneficial.
- Hydrolysis and pyrolysis pathways (and corn-ethanol) are more logical for facilities who have excess steam.
- Gasification pathways are more logical for facilities which buy fossil fuel due to the use of recovered biorefinery heat
- Commercial success is dependent on capital reduction (like grants) and/or unique synergies.
- Feedstocks and markets are the "independent" variables, technology is the dependent variable.
- There may be different technologies of choice depending upon feedstocks and thermal host needs.

Another conclusion is that capital cost should receive more attention from Federal agencies. Neither DOE or USDA have an announced goal in this area and no funding opportunity to date has focused on capital intensity reduction.

References:

(1) Strategic Energy Consultant and Renewable Energy Expert, *email:* bathorp@comcast.net; or phone 804-873-6561.

(2) Presentation at SEKAB, July 2007.

(3) References include web sites for all companies listed, papers they presented at major conferences and private discussion with most

(4) Thorp, Ben, "Paper Industry Must Protect its Status in Cellulosic Innovation", Pulp and Paper, May 2007.

(5) Figure 14, "Advanced Biofuels, Promises, Challenges & Sustainability Issues", Dr. Helena Chum XtL Symposium Montreal Canada, 26 Aug 09

(6) Capital cost based on fall 2009 prices. Operating costs assume $50 per delivered dry ton for slash, spent crops or energy crops.

Preliminary Studies of pH and Char Particle Content On Bio-Oil Aging

Caitlin D. Naske, Dave C. Swalm School of Chemical Engineering, Mississippi State University, Mississippi State, MS

Sarah E. Crosby, Dave C. Swalm School of Chemical Engineering, Mississippi State University, Mississippi State, MS

Andrew McMaster, Dave C. Swalm School of Chemical Engineering, Mississippi State University, Mississippi State, MS

Keisha B. Walters, Dave C. Swalm School of Chemical Engineering, Mississippi State University, Mississippi State, MS

Bio-oil produced by biomass pyrolysis has been shown to degrade over time and at elevated temperatures resulting in increased viscosity that makes processing and handling difficult. It has been theorized that char particles remaining in the bio-oil after pyrolysis and condensation catalyze polymerization reactions leading to the viscosity increases during bio-oil aging. Additionally, pyrolysis bio-oil has a low pH as produced (pH ~2-3) and the effect of the low pH on the polymerization reactions is unknown. In this study, the effects of particulate filtration and pH adjustment are investigated on pyrolysis bio-oil samples produced from ground clear wood cottonwood and pine biomass using an auger reactor under vacuum. The bio-oil control samples were aged up to 21 days at 80 °C and 28 days at 50 °C. For the filtration experiments, pine whole tree fractionated bio-oil was filtered in series by vacuum filtration at 20 µm followed by centrifugal filtration at 0.2 µm and then aged at 80 °C up to 21 days. For the pH adjustment experiments, the pH of pine clear wood fractionated bio-oil was adjusted by the addition of KOH to obtain nominal pH values of 5, 7 and 9 and these pH-adjusted samples were aged at 80 °C up to 14 days. All bio-oil samples were characterized with Fourier transform infrared (FTIR) spectroscopy, gel permeation chromatography (GPC), and combined gas chromatography-mass spectroscopy (GC/MS); select samples were analyzed by light microscopy and electrospray ionization mass spectroscopy (ESI-MS). GPC analysis showed an increase in molecular weight and PDI during aging. The ESI-MS results revealed a polymeric pattern in several of the bio-oil samples, including the aged samples. Light microscopy verified that a large amount of the char particles were removed by centrifugal filtration.

See more of this Session: Integrated Biorefineries for Lignocellulosic Biomass
See more of this Group/Topical: Forest and Plant Bioproducts Division

Production of Fisher-Tropsch liquids from Biomass-derived Syngas

Weihua Deng, Zheng Li, Justinus Satrio and Robert Brown

Center for Sustainable Environmental Technologies (CSET)
Iowa State University
Ames, IA 50011

1. Introduction

Depletion of fossil oil resources, global climate change and national security concerns have made the development of alternative energy increasingly important. Compared to other alternative energy choices, the production of energy and chemicals from biorenewable resources, such as biomass, has attracted increasing research interests in recent years due to the vast potential availability of biomass materials and the availability of existing technologies that potentially enable the production of energy and chemicals in commercial scales in shorter time period. Center for Sustainable Environmental Technologies (CSET) is a research organization at Iowa State University (ISU) that has been working on promoting, developing and demonstrating thermochemical technologies such as gasification and fast pyrolysis, and catalytic upgrading processes, for the production of fuels, chemicals, and energy from biomass. While for many years focuses have been given to gasification and fast pyrolysis technologies for converting biomass into synthesis gas and bio-oil, respectively, in the past two years, CSET researchers have been working on the development of catalytic processes for upgrading syngas and bio-oil into value-added products. This paper is to report the results of research work that has been performed at CSET on the production of liquid fuels from biomass-derived syngas by using Fischer-Tropsch Synthesis (FTS) technology [Dry et al., Anderson et al.]. The work included synthesis and characterization of an FTS catalysts and the demonstration of FTS fuel production from a fixed bed reactor.

One FTS catalyst used for this study is a cobalt-based catalyst which was prepared by using zirconium and γ-alumina as the promoter and support, respectively. Previously, the catalyst has been successfully synthesized, characterized and tested for its performance by using a high pressure slurry reactor constructed at CSET. In order to improve efficiency and liquid selectivity for FTS process, a novel catalyst containing the same active components but supported on SBA-15 based material was developed recently. Since SBA-15 mesoporous silica has large surface area and well defined nanostructures, reaction capacity and shape selectivity of this catalyst is expected to be much better. In this work, the relationships between catalyst property and hydrocarbon fuel production will be thoroughly investigated. The catalysts will be engineered to improve thermal dissipation in the catalyst bed since FTS reaction is highly exothermic. In addition, a comparison between the results of FT synthesis on a slurry reactor and those on a fixed-bed reactor at similar operating conditions will be studied as well. Finally, the effect of impurities in synthesis gas on the catalyst performance will be addressed. This final discussion will relate to the need addressing a compromise between development of robust FTS catalyst and on the development of synthesis gas technology in order to obtain an FTS process that is most sounds techno-economically.

2. **Experimental**

2.1. Catalyst preparation

Catalysts were prepared by means of impregnation with a total metal content of 20wt% Co and 5wt% Zr, respectively. Gamma-alumina and SBA-15 mesoporous silica were used as catalyst supports. The as synthesized catalysts were calcined at 500°C for 6 hours with a temperature increasing rate of 1°C/min form room temperature to the target.

There are two major problems associated with the usage of catalyst powder on the fixed bed reactor: (1) local heating inside catalyst bed resulted from the heat generated from this exothermic reaction, and (2) large pressure drop along catalyst bed which may create gas shortcuts and consequently lower catalyst utilization. To resolve these problems, the catalyst powder was pelletized to make small particles with certain size following the procedure:

- Catalyst powder was pelletized using a Carver press and a set of die to make 0.3 cm thick and ~0.75 cm in diameter pellets. The powder was pressed and held under 5000 lb for 5 minutes.
- Then the pellets were broken down into small pieces.
- A set of sieves was used to separate catalyst particles with different size, and particles with certain size were chosen for this project.
- Selected catalyst was diluted by silica carbide spheres (with similar size to catalyst pellets) and loaded into isothermal zone of the fixed bed reactor.

2.2. Catalyst testing

Figure 1. Fixed bed reactor system constructed at CSET (left: reactor and the right: whole system).

Catalysts were tested on the fixed bed reactor (as shown in Figure 1) constructed at CSET, which is a continuous system with precise flow and pressure control. This system uses circulated oil to provide or remove heat from reaction zone so heat generation of FTS reaction, which probably a perturbation for temperature control if a electrical furnace were used as heating source, will not affect temperature control significantly. Before each FTS reaction, the catalyst was reduced at 350 psi, 330°C under flowing hydrogen (100 mL/min) for 5 hours. Tail gas was sampled and analyzed every one hour throughout the entire reaction period. Liquid products were collected in a condenser after reactionperiod and analyzed by a Varian GC with a FID detector installed, as well as by a GC-MS.

3. **Preliminary results**

A comprehensive investigation of the performance of two above mentioned FTS catalysts under various reaction conditions has not been completed. Some preliminary results obtained so far are discussed in the following.

The activity of homemade FTS catalysts is sensitive to reaction temperature. It was observed that there was almost no activity for the catalysts when reaction temperature was below 210°C. During a temperature range of 220°C to 250°C, FTS catalysts showed good activity as evidenced by a CO conversion between 30% and 70%. On the other hand, total gas selectivity increased significantly as reaction temperature went up indicating less CO was transformed to hydrocarbon fuels. GC analysis of FTS liquid products showed that a broad range of hydrocarbons were generated although quantitative determination of each compound has not been achieved.

Testing of SBS-15 based FTS catalyst under different reaction conditions is underway. By comparing performance of FTS catalysts using different supports, more mechanistic insights into material property-performance relationship can be obtained. The finding will provide useful information in designing the new generation FTS catalyst.

4. **References**

- Dry, M.E., In *Catalysis Science and Technology*; Anderson, J.R., Boudart, M., Eds.; Springer-Veerlag: New York, 1981.
- Anderson, R. B. The Fisher-Tropsch Synthesis; Academic Press: Orlando, FL, 1984.

Development of Comprehensive Models to Predict the Processability of Miscanthus in Biological Conversion

Taiying Zhang, Bourns College of Engineering---Center for Environmental Research & Technology, University of California, Riverside, Riverside, CA

Charles Wyman, College of Engineering-Center for Environmental Research and Technology, University of California, Riverside, Riverside, CA

Bin Yang, CE-Cert, University of California Riverside, Riverside, CA

James Zhang, Mendel Biotechnology Inc., Hayward, CA

Miscanthus, which features with high productivity and high carbohydrate content, has great potential as an energy crop to support large scale production of fuels. Improving our understanding of characterizations of this prolific crop that control enzymatic hydrolysis and how pretreatment alters these characterizations would be a significant step to identify better strategies and opportunities to genetically alter the susceptibility of biomass to deconstruction and achieve very high total sugar yields at low costs. In this study, a high throughput (HTP) tool integrating pretreatment and enzymatic hydrolysis in the same multi-well batch reaction system was employed for screening 80 different Miscanthus species to identify those with lower recalcitrance for sugar release. Then the detailed deconstruction profiles of hemicellulose, cellulose, lignin, and other subcomponents were provided by flowthrough pretreatment at various combinations of time, acid concentration, flow rate, and temperature followed by enzymatic hydrolysis of the pretreated solids. Models were developed based on resulting data to more fully understand key factors to achieve high total sugar yields, and predict the coupled thermochemical and biological deconstruction of Miscanthus, and apply the results to develop much lower cost plant/process combinations.

See more of this Session: Developments in the Pretreatment of Lignocellulosics for Bioconversion II
See more of this Group/Topical: Sustainable Engineering Forum

Effective Hemicellulose Hydrolysis in a Pilot Scale Continuous Pretreatment Reactor; Improving Reactor Design and Understanding Process Performance

Erik Kuhn, National Renewable Energy Laboratory, Golden, CO

Noah D. Weiss, Bioprocess R&D, National Renewable Energy Laboratory, Golden, CO

Nicholas J Nagle, National Renewable Energy Laboratory, Golden, CO

Melvin P Tucker, National Renewable Energy Laboratory, Golden, CO

David Sievers, National Renewable Energy Laboratory, Golden, CO

Richard T Elander, National Renewable Energy Laboratory, Golden, CO

There are many different methods which have been proposed for lignocellulosic biomass pretreatment. A variety of these methods have been shown to exhibit high cellulose and hemicellulose hydrolysis yields in a laboratory scale for pretreatment and the subsequent enzymatic hydrolysis. However, if these methods are going to be used for industrial scale production of cellulosic ethanol, it is necessary to test their performance in scalable, process relevant conditions and equipment. Such tests should be conducted at both a high solids concentration, to increase sugar concentration in resulting fermentation slurries, and in a continuous reactor to determine process performance in an integrated system. We have operated a continuous, horizontal screw reactor for the pretreatment of biomass under acidic conditions and at high solids concentrations. The reactor is capable of pretreating 200 kg per day of dry biomass at elevated temperature and pressures. In this work, we discuss the design of the reactor system as well as the investigation of reactor operating conditions which yield high levels of hemicellulose hydrolysis. Pretreatment experiments using a dilute sulfuric acid pretreatment were carried out to understand the impact of reactor residence time, temperature, and catalyst concentration on the hydrolysis of hemicellulose. Corn stover was pretreated in the reactor under a variety of conditions. Process data indicated significant degradation of hydrolyzed xylan at temperatures above 170°C, and thus lower reaction temperatures were preferred. At reaction temperatures between 150°C and 170°C, more than 80% of the xylan was hydrolyzed with less then 5% of the xylan being degraded to furfural. The hydrolyzed xylan was present as both monomers and oligomers, and an additional hydrolysis step was used to convert the oligomers to monomers. While reaction time, as varied by reactor screw speed, impacted hydrolysis yields, it was difficult to determine an accurate residence time distribution for the system. Overall, the results show that the dilute acid method can be used to effectively pretreat corn stover in a pilot scale continuous reactor and achieve high overall xylan hydrolysis yields.

See more of this Session: Developments in the Pretreatment of Lignocellulosics for Bioconversion II
See more of this Group/Topical: Sustainable Engineering Forum

How Dilute Acid Pretreatment Produces a Highly Digestible Cellulosic Substrate

David K. Johnson, Chemical and Biosciences Center, National Renewable Energy Laboratory, Golden, CO

The structure and chemical composition of plant biomass has evolved to make the cellulose recalcitrant to hydrolysis by cellulase enzymes. Contributing factors to the recalcitrance of lignocellulosic biomass include the epidermal tissue of the plant body, the organization and types of cells present in plants, the difficulty of liquid penetration into the plant cell wall, lignification, the diversity of hemicellulose structure, and cellulose structure. Enzymatic conversion of lignocellulosic biomass requires an effective pretreatment to enable efficient cellulase action. Dilute acid pretreatment substantially increases the accessibility of the cellulose to hydrolytic enzymes and decreases the recalcitrance of the cellulose to hydrolysis. Using a variety of tools, we have developed a better understanding of how the pretreatment process generates a highly digestible cellulosic substrate. Characteristics such as cellulose accessibility, crystallinity, morphology and molecular weight are monitored to ascertain pretreatment effectiveness in addition to other factors including substrate porosity and particle size. Microscopic imaging can indicate changes in lignin and xylan distribution in the plant cell wall that are also important. Higher yields of fermentable sugars, at lower severities, with decreased generation of undesired products and inhibitors, are becoming a reality based on pretreatments designed using the fundamental knowledge being acquired with these developing techniques.

See more of this Session: Developments in the Pretreatment of Lignocellulosics for Bioconversion II
See more of this Group/Topical: Sustainable Engineering Forum

Kinetic Modeling of Lime Pretreatment of Switchgrass

Rocio Sierra, Chemical Engineering, Universidad de los Andes, Bogota, Colombia

Matthew Falls, Chemical Engineering, Texas A&M University, College Station, TX

Mark Holtzapple, Chemical Engineering, Texas A&M University, College Station, TX

Kinetic models applied to oxygen bleaching of paper pulp focus on the degradation of polymers, either lignin or carbohydrates. Traditionally, they separately model two or three different moieties: (1) a rapid, easy-to-remove polymer, (2) medium, and (2) a slow more-recalcitrant polymer. These models were successfully applied to lignin and carbohydrate degradation of switchgrass submitted to oxidative pretreatment with lime at a wide range of conditions: temperature, total pressure and pretreatment time. The models properly fitted experimental data and were used to determine selectivity of pretreatment in two fashions: differential and integral. Through assessment of selectivity, the detrimental effect of pretreatment on carbohydrates at high temperatures and at low lignin content, amply discussed in literature, was ratified.

See more of this Session: Developments in the Pretreatment of Lignocellulosics for Bioconversion II
See more of this Group/Topical: Sustainable Engineering Forum

Cellulose Dissolution in Ionic Liquids: Influence of Ion Structure On β-Glucosidase Stability and Activity

Xinglian Geng, Department of Chemical & Biomolecular Engineering, North Carolina State University, Raleigh, NC

William McDanel, Department of Chemical & Biomolecular Engineering, North Carolina State University, Raleigh, NC

Wesley A. Henderson, Department of Chemical & Biomolecular Engineering, North Carolina State University, Raleigh, NC

Certain ionic liquids are now well known to be effective solvents for cellulose and therefore are of strong interest as solvents for biomass pretreatment and processing. After separating cellulose from raw biomass, enzymatic hydrolysis of the cellulose to D-glucose may be utilized. The influence of the ionic liquid in bulk or trace amounts on the enzymatic activity is not yet well known. Therefore, the effects of four ionic liquids - 1-butyl-3-methylimidazolium chloride ([BMIM]Cl), 1-butyl-3-methylimidazolium acetate ([BMIM]Ac), 1-ethyl-3-methylimidazolium chloride ([EMIM]Cl) and 1-ethyl-3-methylimidazolium acetate ([EMIM]Ac) - on ß-glucosidase stability and activity are compared with a range of inorganic salts with either chloride (Cl^-) or acetate (Ac^- or CH_3COO^-) anions. The ionic liquids changed the enzyme conformation in a very different way from the inorganic salts. Conformational changes of the enzyme did not directly result in the inactivation of the enzyme. When incubated with ß-glucosidase prior to the substrate addition, the four ionic liquids denatured the enzyme irreversibly. With the same anion (CH_3COO^- or Cl^-), the $EMIM^+$ cation was more amicable to the enzyme than the $BMIM^+$ cation. With the same $EMIM^+$ cation, the Cl^- anion was more amicable to the enzyme than CH_3COO^-. However, with the same $BMIM^+$ cation, the Cl^- and CH_3COO^- anions did not result in a significant difference in the enzyme stability. Further, [BMIM]Cl showed stronger inhibition of ß-glucosidase than [EMIM]Ac. These results indicate that the cation in an ionic liquid plays a more important role than the anion in deactivating ß-glucosidase. When the four ionic liquids were instead pre-incubated with the substrates (p-nitrophenyl-ß-D-glucopyranoside or cellobiose) before the enzyme addition, the highest hydrolysis efficiency was obtained with [EMIM]Cl. A higher hydrolysis efficiency was also noted for [BMIM]Cl than for the acetate ionic liquids suggesting that the interactions between the chloride based ionic liquids and the substrates evidently protected the enzyme somewhat from being denatured.

See more of this Session: Developments in the Pretreatment of Lignocellulosics for Bioconversion II
See more of this Group/Topical: Sustainable Engineering Forum

Effect of Biomass Size Partitioning On Pretreatment and Enzyme Requirements Using Response Surface Analysis

Swetha Mahalaxmi, Chemical Engineering, University of Mississippi, University, MS

Geoffrey Sinning, Chemical Engineering, University of Mississippi, University, MS

Clint Williford, Chemical Engineering, University of Mississippi, University, MS

Pretreatment and enzyme hydrolysis are critical steps enabling conversion of biomass to ethanol. They promote separation of hemicellulose and lignin from cellulose and conversion of oligosaccharides to monosaccharides. Partitioning of biomass based on anatomical differences is well accepted as affecting composition and digestibility. Fractions vary in their resistance to milling resulting in various size fractions upon sieving. Milling and sieving could offer simpler partitioning that could be exploited to facilitate pretreatment and more efficient use enzymes for digestion. This would reduce the usage of chemicals, energy and enzymes expended during the production of ethanol. In this work, pre-milled biomass (about 1 inch screen size) was partitioned using 2mm and 1 mm screen sizes, and response surface analysis was performed to evaluate the effect of biomass size partitioning on pretreatment severity and enzyme loading requirements. Pretreatment was conducted, using water at high temperature, as a model pretreatment. Pretreatment time and enzyme loading were the factors investigated using rotatable central composite design for enzymatic digestibility and fermentation of various size fractions of switchgrass.

See more of this Session: Developments in the Pretreatment of Lignocellulosics for Bioconversion II
See more of this Group/Topical: Sustainable Engineering Forum

Fast Pyrolysis for the Production of Bio-Oil From Biomass

Alan H. Zacher, Pacific Northwest National Laboratory, Richland, WA

D.C. Elliott, Pacific Northwest National Laboratory, Richland, WA

To achieve long term sustainability of a biomass-based transport fuel infrastructure will require that alternate feedstocks outside the food chain be developed for conversion to biofuels. One route to this end includes fast pyrolysis of biomass. Fast pyrolysis can produce a 70% yield by weight of a liquid product. The product, called bio-oil, is a fluid mixture of depolymerized and partially decomposed biomass components and water. In the industry, bio-oil has been used directly as a burner fuel, but it also has potential as a quality fuel or chemical feedstock through catalytic hydroprocessing. Pyrolysis also provides a useful mechanism for distributed production of biofuels, and, through collection and transportation of the liquid products to a central processing facility, allows upgrading at a more efficient and cost-effective larger scale. This effort may rely on a number of different sources of biomass feedstock for fast pyrolysis. This research involves the investigation of various types of biomass in a fast pyrolysis and subsequent evaluation of the varying nature of the bio-oil that can be produced. PNNL-SA-66388

See more of this Session: Thermochemical Biorefineries I
See more of this Group/Topical: Forest and Plant Bioproducts Division

High Temperature Steam Gasification of Algae and Switchgrass in a Transport Tube Reactor

R. Bryan Woodruff, Chemical Engineering, University of Colorado at Boulder, Boulder, CO

Paul Lichty, Department of Chemical and Biological Engineering, University of Colorado, Boulder, CO

Carl Bingham, National Renewable Energy Laboratory, Golden, CO

Alan W. Weimer, Department of Chemical and Biological Engineering, University of Colorado, Boulder, CO

Steam gasification of algae and switchgrass to produce synthesis gas (CO, CO2 H2) was performed at high temperature (1273-1673 K) in an externally heated transport tube reactor. High temperature with external heating was investigated because of its applicability to gasification processes using concentrated solar thermal energy. Conventional gasification generally operates at lower temperature and partially oxidizes the biomass to provide the heat of reaction. This either requires an oxygen plant or dilutes the product stream with nitrogen; in either case the heating value of the synthesis gas is lower than the feedstock due to the combustion process. In addition, low temperature often leads to the production of unwanted tars that are costly to remove in downstream processing. Concentrated solar thermal energy offers a solution to both of these issues by providing a low cost, renewable source of high temperature thermal energy. The combination of domestically produced biomass feedstocks with gasification using a renewable heat source offers a viable path to sustainable production of transportation fuels. Algae and switchgrass were studied because of their high land utilization and the growing interest in large scale production using marginal land. The effect of temperature, steam addition, and carrier gas flow rate on conversion efficiency and product distribution was studied in a laboratory scale reactor. Proof of concept experiments were run at NREL's High Flux Solar Furnace to evaluate the performance using concentrated solar thermal energy.

See more of this Session: Thermochemical Biorefineries I
See more of this Group/Topical: Forest and Plant Bioproducts Division

Detailed Tar Analysis and Char Imaging From White Oak Pyrolysis in a Laminar Entrained Flow Reactor

Mark W. Jarvis, National Renewable Energy Laboratory, Golden, CO

Mark R. Nimlos, National Renewable Energy Laboratory, Golden, CO

John W. Daily, Department of Mechanical Engineering, University of Colorado at Boulder, Boulder, CO

Thermochemical conversion of biomass yields a complex slate of light gases, condensable vapors and refractory aromatic tars. However, the detailed reactions and the kinetic parameters that govern their formation are not well known. To address this problem we have designed a novel particle feeder to deliver milled white oak (<180 μm) to our laminar entrained flow reactor (LEFR) for pyrolysis with detection of vapors by molecular beam mass spectrometry (MBMS) and char analysis by transmission electron microscopy (TEM). Two experiments were conducted. A temperature experiment (450-950 °C, ΔT=25 °C) at fixed residence time (1.0 s), and a kinetic experiment varying residence times (0.2-0.8 s, Δt=0.2 s) at five temperature points (500-900 °C, ΔT=100 °C). The spectra from the temperature experiment were analyzed using multivariate curve resolution with optimization by alternating least squares (MCR-ALS). Spectral data from the flow reactor obtained at small increments of temperature over a wide range allows for finer resolution of the changing product slate, and hence the details of the underlying reaction mechanism. We were able to resolve six pure components from the multivariate analysis. The temperature dependence of the pure component scores (relative concentration) are given showing the progression of biomass pyrolysis from primary products (lignin monomers and polysaccharide fragments), to several stages of cracking leading to reduced molecular weight, and finally molecular weight growth to yield benzene, toluene, and polynuclear aromatic hydrocarbons (PAHs). This final pure component is dominant above 850 °C, which correlates with a maximum in the mean particle diameter and circularity as determined from particle imaging. Char particles were embedded and sectioned to provide a three dimensional picture of the morphological changes occurring during the different stages of pyrolysis and gasification. The results show evidence for cell wall, rather than cell lumen, micropore growth into large pores, which could allow entrapment of pyrolysis products and higher concentrations of PAH precursors (C2-C5 species and radicals). These cenospheres begin to rupture after 800 °C, releasing their contents into the gas phase. Since PAHs can only be formed from bimolecular reactions, the entrapment may result in enhanced PAH production. The kinetic experiment provides a thermo-temporal map of tar formation during biomass gasification. Total tars reach a maximum at 600 °C and 0.4 s, while benzene and PAHs increase steadily after 800 °C. MBMS signals for six prominent tar species are used to calibrate MBMS signal to tar yield (mg tar per g white oak). This information will be used to improve a semi-detailed reaction mechanism of biomass gasification for use in CFD modeling.

See more of this Session: Thermochemical Biorefineries I
See more of this Group/Topical: Forest and Plant Bioproducts Division

Effect of Particle Size On Tar Formation in a Bench-Scale Gasifier Using Woody Biomass

Katherine R. Gaston, National Bioenergy Center, National Renewable Energy Laboratory, Golden, CO

Mark W. Jarvis, National Bioenergy Center, National Renewable Energy Laboratory, Golden, CO

Mark Nimlos, National Bioenergy Center, National Renewable Energy Laboratory, Golden, CO

Kristin Smith, National Bioenergy Center, National Renewable Energy Laboratory, Golden, CO

We employed a four-inch diameter fluidized bed reactor in combination with NDIR/TCD and molecular beam mass spectrometer to study the effect of biomass particle size on the yield of light gas and tar compounds during gasification. White oak was milled into four sphere sizes with diameters of 0.25, 0.5, 0.7, and 1.0 inch. Reactor bed temperatures of 700, 800, and 900°C were used. As expected, temperature had the greatest effect on tar production with yields increasing sharply after 800°C. Furthermore, the trend for the total tar yield as a function of particle size is complex. At 900°C, we observed a peak for 0.5 in. spheres; at 800°C total tar yield was relatively flat; at 700°C the 1 in. spheres produced relatively more total tar. Interestingly, the primary peaks for the PAH compounds, m/z 78 (benzene) and m/z 128 (naphthalene), increase with decreasing particle size, and the peak yields for the 0.5 in. spheres are still observed at 900°C. These initial observations suggest that PAH production is exacerbated by biomass sample geometries with a high surface area to volume ratio, while the total quantity of light gases produced at high temperature remains constant. The lower internal heating rates experienced by large diameter spheres favor an increase in total tar production at the expense of PAH production. The gas pulse observed in both the NDIR/TCD and the MBMS for the smaller, 0.25 in. spheres was on the order of 20 seconds and, for the 1 in. spheres, was between 2 and 3 minutes. This also shows that the smaller spheres are more rapidly heated than the larger ones and hence are subjected to more severe pyrolysis throughout the entire gasification process.

See more of this Session: Thermochemical Biorefineries I
See more of this Group/Topical: Forest and Plant Bioproducts Division

Rational Design of Tar Reforming Catalysts

Kim Magrini, National Renewable Energy Laboratory, Golden, CO

Whitney Jablonski, National Bioenergy Center, National Renewable Energy Laboratory, Golden, CO

Matthew M. Yung, National Bioenergy Center, National Renewable Energy Laboratory, Golden, CO

Yves Parent, National Bioenergy Center, National Renewable Energy Laboratory, Golden, CO

The efficient thermochemical conversion of biomass to produce liquid fuels using indirect gasification requires the use of a catalytic steam reforming process to remove and reform tars to produce clean syngas which additionally maximizes the use of the biomass carbon and hydrogen content. The catalysts used in the steam reforming reactor can comprise a significant portion of the overall process cost to produce biofuels. In order to reduce the catalyst cost and improve the overall fuel synthesis economics, highly active and stable catalysts are required. To achieve this goal, we are using rational catalyst design to understand, design, and synthesize robust, tailor-made catalysts capable of handling biomass-derived impurities such as tars, polyaromatic hydrocarbons (PAHs), and H2S. Results of catalyst design and reforming performance are discussed.

See more of this Session: Thermochemical Biorefineries I
See more of this Group/Topical: Forest and Plant Bioproducts Division

Visualization of Combustion of Solid Fuel Particles in a Drop Tube Furnace

Fengming Huang, Chemical Engineering, Auburn University, Auburn University, AL

Steve R. Duke, Chemical Engineering, Auburn University, Auburn University, AL

A laboratory-scale drop tube furnace was designed to maximize visualization capability for combustion of fuel particles and droplets in combustion environments with a high degree of control of temperature, pressure, and gas composition. The furnace operates at temperatures up to 1,200°C and it has three independent temperature zones. The drop tube retort is transparent quartz with a diameter of 6 cm and a length of 80 cm; three vertical 25 cm rectangular windows in the furnace casing allow visualization access to particles as they combust at various distances from a water-cooled injection nozzle.

Results will be presented for single particle combustion of samples of pulverized coal, wood chips, and switch grass. The combustion of individual particles in the furnace was visualized with a high-speed camera and a high magnification lens system. Video provides qualitative comparison of combustion events, and still images were analyzed to measure surface temperature distributions and burn properties for the particles. Results will be presented for 100 micron particles undergoing combustion at 900°C. The fuels, particle sizes, and operating conditions of the furnace were selected to evaluate effectiveness of these alternative fuels in comparison to coal for application in cement production.

See more of this Session: Thermochemical Biorefineries I
See more of this Group/Topical: Forest and Plant Bioproducts Division

Two-Stage Pretreatment of Miscanthus with Alkaline Peroxide and Electrolyzed Water for Ethanol Production

Atilio de Frias, Department of Agricultural and Biological Engineering, Energy Biosciences Institute, University of Illinois at Urbana-Champaign, Urbana, IL

Xiaojuan Wang, Department of Food Science and Human Nutrition at UIUC, Chemical Machinery at DUT, University of Illinois at Urbana-Champaign (UIUC) and Dalian University of Technology (DUT), Urbana, IL

Hao Feng, Department of Food Science and Human Nutrition, Energy Biosciences Institute, University of Illinois at Urbana-Champaign, Urbana, IL

Miscanthus x Giganteus, a C4 perennial grass, has been proven in recent studies to be twice as productive as switchgrass with regard to yields of biomass produced per acre (Long et al. 2004). Miscanthus' low nutrient and water requirements, as well as its productivity in marginal lands, make it an ideal energy crop. As a relatively new energy crop, the deconstruction of Miscanthus for bio-ethanol production is of interest.

The Miscanthus samples used in these experiments were grounded to pass a 50-70 mesh sieve, and are composed of 49.0% glucan, 22.5% xylan, 1.2% arabinan, 1.6% galactan and 21.6% lignin. The chemical composition analysis was performed following the NREL (National Renewable Energy Laboratory) methods.

Miscanthus was subjected to alkaline peroxide and electrolyzed water two-stage pretreatment. The electrolyzed water was produced by electrolysis of ordinary tap water containing low concentration of sodium chloride (0.1%). In the first stage, Miscanthus was soaked in an alkaline peroxide solution (pH=11.5) at five different concentrations (1-5% w/w) with a 5% solids loading at 50 C for 24 h, to partially remove hemicellulose and lignin. In the second stage, the cellulose-rich solids were pretreated with electrolyzed water at 121 C for 15 min, with a 7% solids loading.

After the second pretreatment, the slurries (6% solids) were hydrolyzed with 15 FPU/g glucan of cellulase and 40 U/g glucan of b-glucosidase at 50 C for 72 h. Following the enzymatic hydrolysis, sugar concentrations were determined by HPLC. Finally, a fermentation study was performed at 30 C and pH=5.5 for 72 h using Saccharomyces Cerevisiae to determine ethanol yield.

An enzymatic digestibility of up to 96% was achieved with the proposed two-stage pretreatment method, with relatively mild temperatures and shortened enzymatic hydrolysis times. With no detoxification performed, the hydrolysate produced from the two-stage process exhibited improved fermentability.

Keywords: Miscanthus, electrolyzed water, pretreatment, enzymatic digestibility

References

E. Heaton, T. Voight and S.P Long. A quantitative review comparing the yields of two candidate C4 perennial biomass crops in relation to nitrogen, temperature and water, Biomass and Bioenergy 27 (2004), pp. 21-30

See more of this Session: Developments in the Pretreatment of Lignocellulosics for Bioconversion III
See more of this Group/Topical: Sustainable Engineering Forum

Hydrothermal Pretreatment of Switchgrass for Ethanol Production

Sandeep Kumar, Chemical Engineering, Auburn University, Auburn, AL

Urvi D. Kothari, Chemical Engineering, Auburn University, Auburn, AL

Lingzhao Kong, Chemical Engineering, Auburn University, Auburn, AL

Y. Y. Lee, Chemical Engineering, Auburn University, Auburn, AL

Ram B. Gupta, Chemical Engineering, Auburn University, Auburn, AL

Switchgrass is emerging as a potential lignocellulosic biomass for ethanol production. Unfortunately, switchgrass has a very low enzymatic reactivity. Therefore, pretreatment of switchgrass is viewed as a critical step to make the cellulose accessible to enzymes and for an adequate yield of fermentable sugars during enzymatic hydrolysis. Hydrothermal pretreatment of biomass has attracted much attention as a non-toxic, environmentally benign, and inexpensive method. The process mainly removes lignin and resulting structural changes improve the accessibility and hydrolysis of cellulose. The increase of ionization constant of water with temperature and formation of organic acids during pretreatment process causes the reaction medium to become more acidic (pH <4) and enhances the hydrolysis of majority of hemicelluloses. The hydrothermal pretreatments of switchgrass and corn stover were conducted in a semi-continuous reactor to enhance and optimize the enzymatic reactivity. The study was aimed to maximize the retention of hemicelluloses in solid fraction and increase the removal of lignin during such treatments. The effects of pretreatment temperature (150-200°C), pressure 34-138 bar, and addition of K_2CO_3 (0-0.90 wt.%) were examined. The enzymatic reactivity of the pretreated substrate was investigated with cellulase and β-glucosidase enzymes. In addition, the changes in crystallinity, surface area, and surface morphology were examined by X-ray diffraction, BET surface area measurement, and electron microcopy, respectively.

See more of this Session: Developments in the Pretreatment of Lignocellulosics for Bioconversion III
See more of this Group/Topical: Sustainable Engineering Forum

Solvent Fractionation of Biomass: Applications of Switchgrass Pretreatment for Biofuels

C.J. O'Lenick, Forest Products Center, University of Tennessee, Knoxville, TN

Joseph J. Bozell, Forest Products Center, University of Tennessee, Knoxville, TN

Timothy G. Rials, Forest Products Center, University of Tennessee, Knoxville, TN

Rene Buchinger, University of Applied Sciences Salzburg, Salzburg, Austria

Anton Astner, University of Applied Sciences Salzburg, Salzburg, Austria

Stuart Black, National Renewable Energy Laboratory, Golden, CO

Stacy A. Warwick, Forest Products Center, University of Tennessee, Knoxville, TN

Interest in recovering cellulose from biomass has grown steadily over the last few years as cellulose can be enzymatically hydrolyzed to sugars and fermented to ethanol. Under the Energy Independence and Security Act of 2007, refiners must produce 16 billion gallons of cellulosic ethanol per year by the year 2022. Switchgrass is a very promising lignocellulosic source for this ethanol. Switchgrass is a versatile and adaptable plant, as it can grow in different weather conditions. It is a perennial and can grow in poor soils that cannot support any food crops. It is not a food source so it will not compete for food crops. We are investigating solvent fractionation (organosolv processing) to isolate the required cellulose from switchgrass. Solvent fractionation is a process of choice for pretreatment because it is suitable for use with several different biomass feedstocks, giving favorable separations, easy isolation of products after fractionation, and offers recovery of each component in a high yield and purity amenable to conversion to other chemicals. Fractionation technology developed at the National Renewable Energy Laboratory is being used at the University of Tennessee for the separation of switchgrass into its primary components. The fractionation employs a mixture of organic solvents and water to separate switchgrass into cellulose, lignin and hemicellulose for the production of fuels and chemicals. Fractionation is carried out by adding biomass, a ternary solvent mixture, and a sulfuric acid catalyst to a 3.5 L, 3" bore, pressurized, Hastelloy flow-through reactor, controlled via LabVIEW and operating at three temperatures: 120 deg C, 140 deg C, and 160 deg C. The recovered solvent is subjected to a phase separation, giving an organic phase containing lignin and an aqueous phase containing hemicellulose. The cellulose fraction is obtained as a solid in an average yield of 38.6% by weight (7 runs). Lignin yield at an average of 6.3% by weight (9 runs) is fairly constant and independent of temperature and pressure. 2D NMR data used to determine structural changes in lignin as a function of separation conditions will be presented.

See more of this Session: Developments in the Pretreatment of Lignocellulosics for Bioconversion III
See more of this Group/Topical: Sustainable Engineering Forum

High Value Products Derived From Biological Pretreatment

Deepak Singh, Department of Biological Systems Engineering, Washington State University, Pullman, WA

Xiaochen Yu, Department of Biological Systems Engineering, Washington State University, Pullman, WA

Sulin Chen, Department of Biological Systems Engineering, Washington State University, Pullman, WA

Biological pretreatment for breaking down lignin and hemicellulose has been identified as one of the research priorities as an alternative to thermochemical pretreatment processes of lignocellulosic biomass. In destruction of biomass, microorganisms have the ability to liberate compounds that could be recovered as co-products. This study depicts the biological pretreatment process in combination with mild thermochemical treatment to hydrolyze wheat straw, releasing high value products as well as fermentable sugars. Phanerochaete chrysosporium was used to pretreat the straw for up to three weeks at 37oC. Tween 80, lactic acid, and H2O2 were used in the sugar production analysis, and 0.1 % Tween 80 treated straw showed the best yield. It is also important to note that after mild acid hydrolysis (0.2% H2SO4) the total sugar recovery was also stable in the sample supplemented with Tween 80. To enhance hydrolysis, samples were subjected to mild acid treatment at 121oC/20 min. A total sugar content of approximately 18 g/L was detected in the first week sample, where the control gave approximately 6 g/L. However, the enzymatic hydrolysis gave no significant change in sugar level compared to the control. Therefore, the experiment suggests that the mild acid treatment may be a useful method when combined with biological pretreatment to extract the hydrolysable sugars. Decaying of biomass was observed by scanning electron microscopy (SEM). Fourier Transform Infrared (FTIR) spectroscopy showed a distinct absorbance at 1500-1800 cm-1, indicating the deposition of C=O stretching compounds, 1497-1530 cm-1 the aromatic skeletal, and 1383-1360 cm-1 for aromatic methoxy stretching. The result was further verified by fast pyrolysis GC/MS. The analysis of each individual peak after pyrolysis GC/MS was done and showed a clear increasing trend (approximately 3%) in the concentration of venylguaicyl and syringol, with the biomass incubation time was noted. This indicates that during biological pretreatment, lignin monomers such as venylguaicyl and syringol units can be extracted in high amounts after a mild thermal treatment, which can be further used as fuel compounds. Interestingly, a compound which can be used for the development of anticancer compounds was discovered among the new compounds after biological pretreatment process and is currently under characterization. Overall, this approach of biomass processing is a new way to utilize the biomass, not only for obtaining fermentable sugars, but also for the extraction of valuable bioproducts.

See more of this Session: Developments in the Pretreatment of Lignocellulosics for Bioconversion III
See more of this Group/Topical: Sustainable Engineering Forum

Ionic Liquid Pretreatement of Lignocellulosic Material for Enhancement of Enzyme Hydrolysis

Indira Priya Samayam, Chemical Engineering, University Of Toledo, Toledo, OH

Constance Schall, Chemical Engineering, University Of Toledo, Toledo, OH

Sasidhar Varanasi, Chemical Engineering, University Of Toledo, Toledo, OH

Depleting fossil fuels and the demand for energy has necessitated the urge for an alternative renewable energy and lignocellulosic material offer a sustainable and renewable source for the production of biofuel.

A novel ionic liquid pretreatment technology has been developed to reduce the robustness of cellulose and make it amenable to the cellulase enzymes. The ionic liquid pretreatment of lignocellulosic biomass produces amorphous or partially crystalline cellulose and enhances the saccharification rates of biomass sugars. The high crystallinity of cellulose poses an impediment to hydrolysis enzymes for the cellulosic portion of biomass. Different pretreatment conditions have been explored to produce amorphous cellulose to increase accessible surfaces to the enzymes used in saccharification. Variable enzyme mixes have been used to study the saccharification kinetics in an attempt to explore a simplified enzyme component mixes. Cellulose and biomass pretreated with ionic liquid have been characterized to assess the crystallinity and elucidate the accessibility of cellulase enzymes to biomass. Saccharification results for pretreated biomass hydrolyzed with different cellulase enzymes are compared and also evaluated using simplified enzyme mixtures.

See more of this Session: Developments in the Pretreatment of Lignocellulosics for Bioconversion III
See more of this Group/Topical: Sustainable Engineering Forum

The Kinetics of Reversion Reactions During Acid Pretreatment of Biomass

Mark R. Nimlos, National Renewable Energy Laboratory, Golden, CO

Heidi M. Pilath, National Renewable Energy Laboratory, Golden, CO

Michael E. Himmel, Chemical and Bioscience Center, National Renewable Energy Laboratory, Golden, CO

David K. Johnson, National Bioenergy Center, National Renewable Energy Laboratory, Golden, CO

Reversion reactions of the sugars released during acid pretreatment have been studied using a microwave-heated robotic reactor system. These reactions represent a potential loss of sugars during biofuels processes and knowledge of the kinetics of these reactions is necessary to optimize pretreatment. Towards that end the rates of reactions of monomer sugars to form dimers have been studied. Though xylose dimerization is difficult to study due to the lack of standards, reversion reactions of glucose are tractable. The rates for glucose were measured and these have been used to predict rates for xylose, which is more prevalent during pretreatment. The mechanisms for forming disaccharides involve dehydration of one monomer, forming a carbo-cation at C1 and addition by the hydroxyl group of another sugar molecule. Thus, reversion reactions necessarily result in dimers link at the C1 position of one of the sugars. In addition, we have observed that at temperatures above 140 °C equilibrium is rapidly established between glucose and the resulting disaccharides. The kinetics of glucose reversion reactions are measured and modeled and are used to predict rates for xylose. The impact of these rates on optimal pretreatment conditions will be discussed.

See more of this Session: Developments in the Pretreatment of Lignocellulosics for Bioconversion III
See more of this Group/Topical: Sustainable Engineering Forum

Simultaneous Infrastructure Investment and Operation Decisions for Bio-Refinery Networks

Jinkyung Kim, School of Chemical & Biomolecular Engineering, Georgia Institute of Technology, Atlanta, GA

Matthew J. Realff, School of Chemical & Biomolecular Engineering, Georgia Institute of Technology, Atlanta, GA

Jay H. Lee, School of Chemical & Biomolecular Engineering, Georgia Institute of Technology, Atlanta, GA

Craig Whittaker, Research & Development, Weyerhaeuser NR

Ludwig Furtner, Research & Development, Weyerhaeuser NR

Amar Neogi, Research & Development, Weyerhaeuser NR

There is a growing interest and technological advances being made in converting biomasses into bio-fuels. Even though we are still years away from its full commercialization, it is important to begin to assess the economic feasibility of various emerging technologies and evaluate the advantages and disadvantages of different design and logistics options. In this presentation, we propose an optimization model that enables the selection of fuel conversion technologies, processing capacities and locations, and the logistics of transportation from forestry resources to conversion and from conversion to final markets. A mixed integer linear program (MILP) model is built and implemented in a commercial software package (GAMS) using databases built in Excel.

The MILP solves for (1) the optimal numbers, locations, and sizes of various types of processing plants and (2) the amounts of biomass, intermediate products, and final products to be transported between the selected locations, with the goal of maximizing the overall profit under present constraints. It also outputs the cost and profit data associated the selected network in a convenient form for analysis. The model is tested using a industry-representative data set that contains information on the existing wood resources, final product market locations and demands, candidate locations and sizes for different types of processing plants, and costs associated with the various processing units and transportation of materials, within the Southeastern region of the United States. We investigate: 1) Which parameters have major effect on the overall economics, and 2) benefits of going to more distributed types of processing network, in terms of the overall economics and the robustness to demand variations.

See more of this Session: Thermochemical Biorefineries II
See more of this Group/Topical: Forest and Plant Bioproducts Division

INTEGRATING BIOMASS TORREFACTION WITH THERMO-CHEMICAL CONVERSION PROCESSES

Dr. Sudhagar Mani, Faculty of Engineering, University of Georgia, Athens, GA
Email: smani@engr.uga.edu

Introduction

Agricultural and forest residues are the major feedstock for a biorefinery that can process and convert them into valuable chemicals, heat, fuels and materials. Biomass in its original state have wide range of moisture content (25-60%), low bulk density (60-100 kg/m^3) and low energy density (8-14 MJ/kg) and is difficult to transport, handle, store and feed into the existing combustion and gasification systems (Sokhansanj et al., 2002; Mani et al., 2006b). The uneven and high moisture content of biomass feedstock make it difficult to transport and store for a long time before its use in the biorefinery. Energy required for grinding of biomass is very high to produce a powdered biomass supply at the throat of the conversion chamber (Mani et al. 2004). In addition, gasification of lignocellulosic biomass with high volatiles and moisture content yields heavy tar formation that makes the process more expensive and harder to operate, .e.g. tar contamination in catalysts used in downstream conversion of syngas into methanol and ethanol (Dayton, 2002; Boerrigter et al., 2005). Major power companies are interested in co-firing of biomass with coal, but face several operating issues due to imbalance in the energy density of coal and biomass, low burning rate and low flame temperature. Biomass also creates additional problem due to poor flowability, poor blending and reduces the thermal efficiency and capacity of existing boiler units (Tillman, 2000). Pyrolysis of lignocellulosic biomass yields high molecular weight organic liquid called bio-oil and a carbon rich material called bio-char. Bio-oil is recognized as a major feedstock for producing chemicals, fuels and heat. However, application and use of bio-oil have several technical challenges due to low pH, high oxygen and water contents, high molecular weight tar compounds and high viscosity. Several of the above problems on the thermo-chemical conversion biomass were partly or mostly originated from feedstock properties. Thus, thermal pretreatment of biomass or torrefaction process can provide a promising solution for thermo-chemical conversion technologies.

Torrefaction is a process of thermal degradation of biomass at relatively low temperatures (200-300°C) under anoxic conditions resulting in an energy dense, stable char or coal-like material that also eliminates or minimizes coke forming precursors, such as acetic acid. Removal of these coke forming precursors could significantly improve catalytic upgrading of the bio-oil generated by pyrolysis. It is anticipated that integrating biomass torrefaction with thermo-chemical conversion processes will not only improve the energy density and quality of bio-crude oil/syngas, but also minimizes preprocessing and storage costs. Co-firing torrefied biomass with coal can be possible without compromising the energy efficiency and plant production capacity. This paper present the recent progress on the torrefaction characteristics of forest biomass and the effect of torrefaction on the pyrolysis and gasification processes studied using a thermal gravimetric analyzer with mass spectrometry (TG-MS).

Materials & Methods

Materials

Southern pine forest residue was selected as a biomass feedstock for this study. Large volume of forest residue was collected from Macon, GA and was frozen for the torrefaction experiment. The physical properties of forest residues such as moisture content, mass distribution, bulk density, particle density and particle size were determined based on the procedures used by Mani et al. (2004 & 2006a). The proximate and ultimate analyses and heating value of the biomass were determined as per standard procedures. To study the effect of torrefaction with various thermo-chemical conversion processes, southern pine chips were used.

Torrefaction Experiment

A batch torrefaction reactor (Figure 1a) with a capacity to hold 3-5 lbs of biomass samples was used to study the torrefaction characteristics of forest residue. The batch reactor has three ports as 1) for temperature sensor; 2) for carrier gas inlet and 3) for carrier gas outlet. We used nitrogen as a carrier gas for all the torrefaction experiment. For each torrefaction experiment, about one lb (0.454 kg) of forest residue was placed in the cast iron container and was sealed and placed in the temperature controlled oven. The oven temperature and sample temperature data were data logged in the computer and the temperature inside the reactor was controlled accurately. Figure 2b shows a typical temperature profile during torrefaction of forest residues. A sweep gas flow rate of 2 L/min of nitrogen was used to maintain anoxic condition inside the reactor and the exhaust gas line was passed through a series of condensers to collect all the condensable compounds generated from the torrefaction experiment. The non-condensable gases were passed though a water column to remove any dirt and were collected using gas bags for further analyses. Once the torrefaction temperature and holding time were reached, the sample will be cooled using liquid nitrogen or by forced convection until the sample reaches close to room temperature. Figure 2(b) shows the typical time –temperature profile of a batch torrefaction reactor. Then, the torrefied samples were removed from the reactor, weighed and stored for further analysis. The condensed liquids were collected in sample tubes for GC-MS analysis.

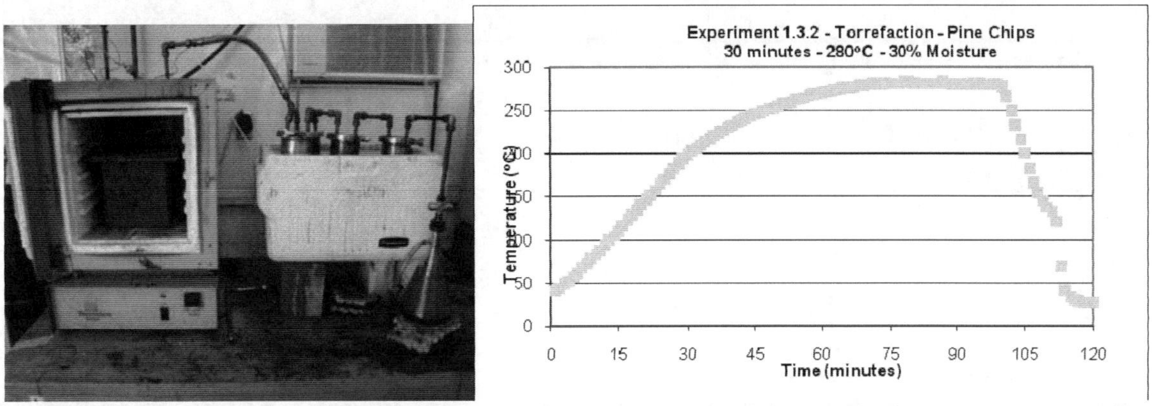

Figure 1. Schematic diagram of a batch torrefaction reactor (a) with a typical temperature profile (b)

The quality of torrefied biomass depends on the number of process variables, which are chemical composition of the biomass, particle size, moisture content, reaction temperature, reaction time, heating rate and types of carrier gases. Among various parameters, torrefaction temperature has a significant effect on the yield and composition of biomass. Therefore three different torrefaction temperatures: 220, 250 & 280°C were selected for this study. Torrefied biomass collected at the end of each experiment was analyzed for chemical compositions, heating value, and structural characteristics of biomass. Finally, the process variables responsible for producing high energy density torrefied biomass (biocoal) were optimized.

Integration of torrefaction with thermo-chemical conversion processes

In order to study the effect of biomass torrefaction with thermo-chemical conversion processes such as combustion, gasification and pyrolysis processes, a Thermal Gravimetric Analyzer with Mass Spectrometry was used. Southern pine chips were ground using a knife mill and passed through 0.25 mesh size was used for this study. For each experiment, about 15 mg of sample was torrefied at four temperatures (225, 250, 275 & 300°C) and the solid samples were quenched with liquid nitrogen. Then, the samples were used for combustion, gasification and pyrolysis experiments respectively using air, partial oxygen mixture and nitrogen as sweep gases. For each type of thermo-chemical conversion processes, mass degradation rate and the off-gas compositions monitored using selective ion were recorded and analyzed. Each test combinations were repeated trice.

RESULTS & DISCUSSIONS

Forest residues characterization

Forest residues were fractionated into clean chips, bark, small stems, leaves, and small particles. More than half (51%) of the samples were chips, followed by 19% of small stems and 13% of bark. Small stems and particles contributed about 8% and 9% respectively. Most of the small particles were pine needles ground up during the chipping operation. The particle size distribution of each mass fraction is given in Table 1. The average bulk density of the whole forest residues was (on dry basis) 149.17 kg/m^3 (SD = 4.47), compared to sawdust of 142.2 kg/m^3 and clean pine chips of 209.8 kg/m^3 (Gigler et al. 2000a). The lower bulk density of forest residues was mainly attributed to uneven particle size distribution of small stems and leaves.

Table 1. Particle Size Distribution of Components of Forest Residues

Residue fractions	Thickness[a] (mm)	Length[a] (mm)	Width[a] (mm)
Chips	6.3 (0.8)	26.1 (8.3)	18.5 (0.0)
Barks	4.4 (1.2)	32.0 (0.9)	15.0 (0.8)
Small stems	4.7 (0.5)	124.9 (93.3)	4.6 (0.2)
Leaves	0.5 (0.1)	93.0[b] (28.2)	___
Small particles[c]	<4.8		

[a] Numbers in the parenthesis are standard deviation with five replications.
[b] The length of the leaves ranges from 36.1 to 124.1 mm.
[c] The size of small particles represents particles that are smaller than 4.8 mm screen size.

Biomass Torrefaction Experimental Data

Figure 2 shows the example of torrefied products obtained from each experiment. Figure 3 shows the yield of torrefied biomass (in dry basis) at various temperatures and initial residue moisture content with 30 min residence time. In overall, lower torrefaction time in the batch reactor results in higher yield as the solid hydrolysis of hemiellulose takes longer time to decompose into vapors. Among the variables studied, torrefaction temperature and residence time had significant yield on the total yield of torrefied biomass. Torrefaction temperature had a significant effect on the torrefied biomass yield, chemical compositions and the heating value of biomass. The results were comparable with experimental data obtained by Prins et al. (2006a). Tables 3 & 4 show the changes in chemical composition of forest residue at various torrefaction temperatures. It was evident that an increase in torrefaction temperature increased the heating value of biomass to a level closer to that of lignite or low grade bituminous coal. A further increase in torrefaction temperature will result in low yield of biomass and an increase in the volatile oil (bio-oil) yield. Therefore, a torrefaction temperature of 280°C is optimal for forest biomass based on the heating value and the yield of biomass obtained from this process.

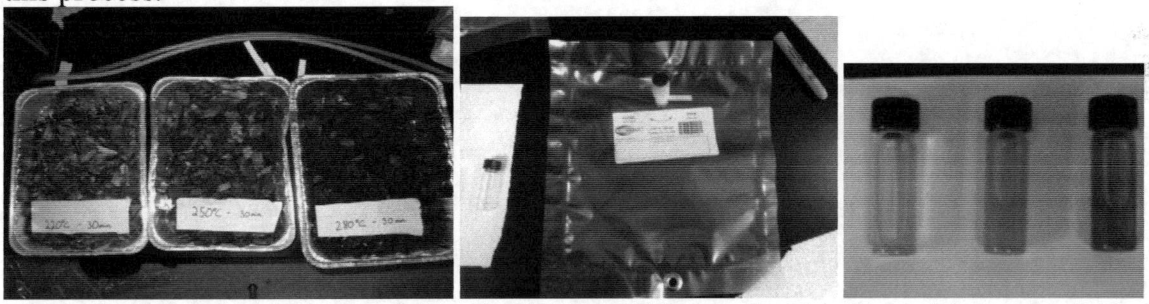

Figure 2 shows the products of torrefied biomass (a) torrefied biomass; (b) non-condensable gases and (c) condensable liquids

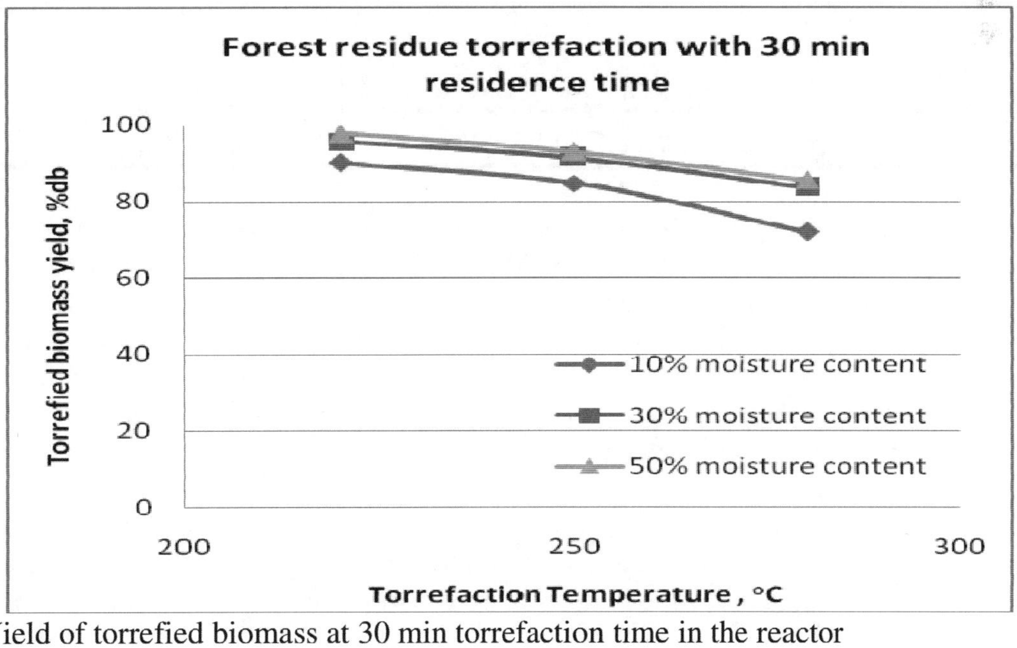

Figure 3. Yield of torrefied biomass at 30 min torrefaction time in the reactor

High initial moisture content of biomass is not preferred as more time and energy required removing moisture from the biomass before starting the torrefaction process. If low temperature gas sources are available in the power plant, this process can be integrated to the power plant for efficient utilization of low heat energy sources such as exhaust gases. A significant reduction in the composition of hemi-cellulose was noted with an increase in torrefaction temperature indicating that hemi-cellulose is thermally hydrolyzed during torrefaction (data not shown). Some degree of cellulose was also decomposed during torrefaction and the rate of cellulose decomposition increased with increase in torrefaction temperature.

Table 3. Elemental analysis of torrefied forest residue samples

Torrefaction temperature °C	**C (%)**	**H (%)**	**O (%)**	**N (%)**	**S (%)**
Non-torrefied	45.3	5.9	48.5	0.3	0.1
220	49.4	5.5	44.1	0.3	0.0
250	50.5	5.4	41.8	0.4	0.0
280	56.4	5.4	36.1	0.9	0.1

Table 4 Proximate analysis of torrefied forest residue samples

Torrefaction temperature °C	**Moisture (%)**	**Volatile C (% wt)**	**Fixed C (% wt)**	**Heating value (Btu/lb)**	**Ash (%)**
Non-torrefied	10	75.3	16.3	7,774	0.67
220	3.2	76.8	19.1	8,474	1.1
250	2.3	74.9	20.6	9,376	1.2
280	2.1	70.8	25.6	10,167	1.4

Figure 4 shows the concentration of various chemicals present in the non-condensable portion of gases collected during torrefaction of forest residues at 280°C. Three major compounds: carbon monoxide, carbon dioxide and methane were emitted in large quantities during the early stage of torrefaction and their concentrations were decreased as the residence time increases. Other compounds such as hydrogen, propylene, ethane, ethylene, n-butane and methyl acetylene were also present in the gases at low concentrations. The rate of emission of above compounds varies significantly as the torrefaction temperature changes and can be potential used as a carrier gas during torrefaction process instead of nitrogen as an inert gas.

Figure 5 shows the partial concentration of condensable liquids collected during torrefaction of biomass at various residence times. Although several hemi-cellulose and extractives degraded compounds were noticed from the GC-MS data, higher concentrations of alpha-pinene, camphene, beta-pinene and beta-phellandrene were present.

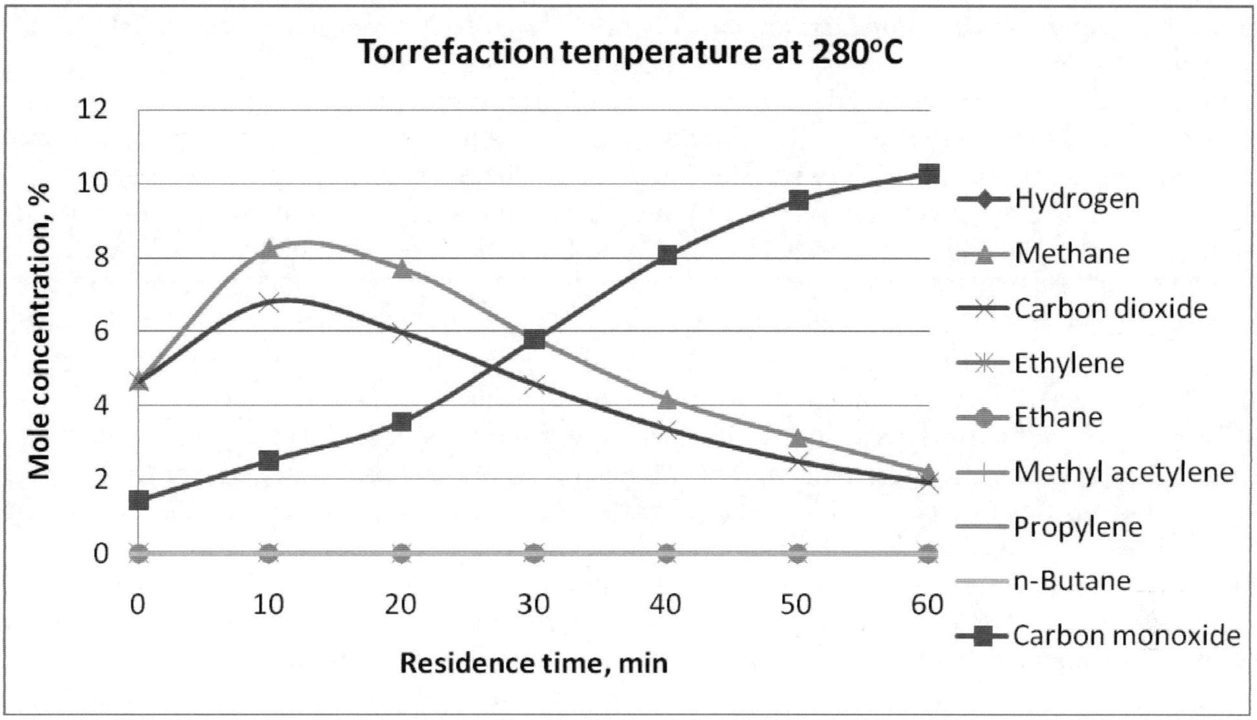

Figure 4. Concentration of various compounds present in the non-condensable gases generated during torrefaction of forest residues.

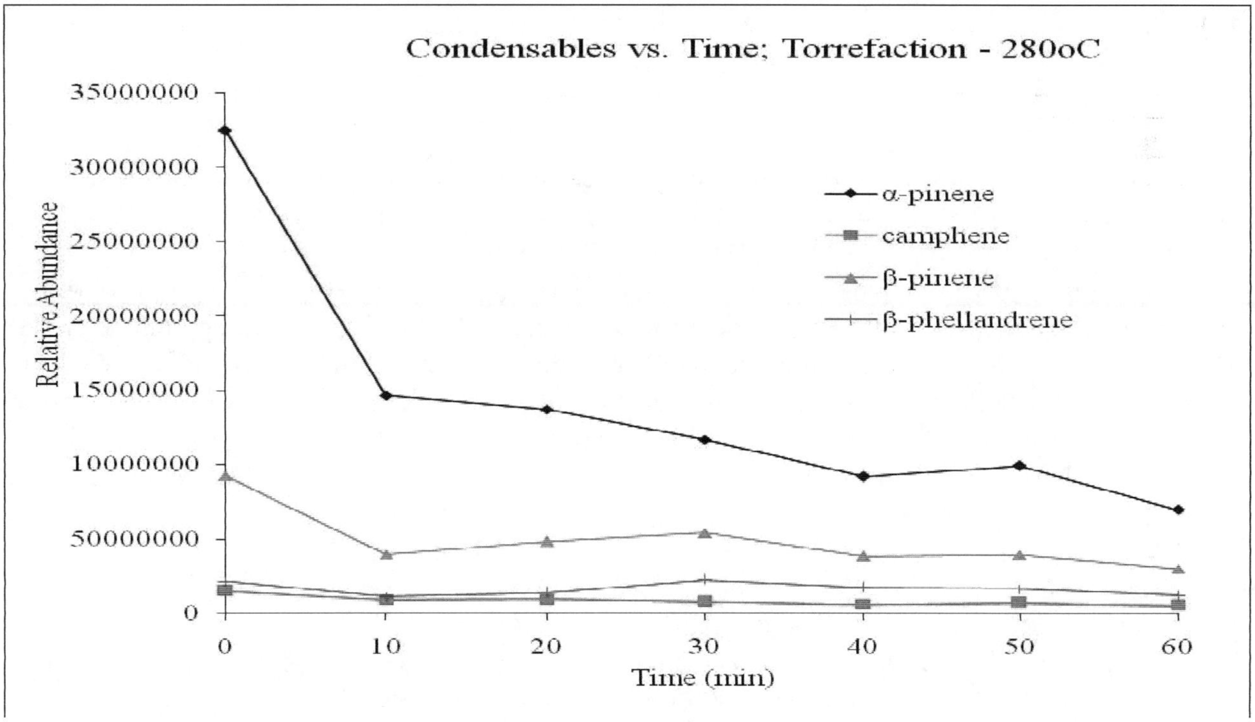

Figure 5. Qualitative composition of major chemicals present in the condensable liquids collected at various torrefaction times.

Integrating Torrefaction with Combustion, Gasification & Pyrolysis Processes (preliminary data)

Combustion: Based on the mass decay data from TG-MS experiment, both the control and torrefied biomass was more reactive than coal. The rate of mass loss was higher for both control and torrefied biomass compared to coal. As the torrefaction temperature of biomass increases, little or no effect on the rate of mass loss was noticed. This indicates the there were no significant difference among the control and torrefied biomass burning rate. The preliminary data from the combustion study will be useful for conducting the lab scale combustion study on both torrefied biomass and coal.

Gasification: Mass spectrometry (MS) data collected using selective ion monitoring from torrefaction with gasification experiment showed that there were significant reduction in carbon dioxide and acetol production when pine was torrefied at 275°C for 20 min compared to untreated (control) biomass. Furthermore, a marked reduction in peak acetic acid production relative to baseline signal was observed for torrefied samples compared to control biomass. There was a statistically significant change in the amount of guiacol (One of the critical pre-cursor compounds for tar formation during biomass gasification) detected in one sample (torrefied at 225°C) compared to the control. The results were partially comparable with results obtained from Prins et al. (2006b). Further research on this topic is required to critically evaluate the effect of torrefacion on the gasification of biomass.

Pyrolysis: MS data from torrefaction with pyrolysis experiment indicated that higher torrefaction temperatures resulted in a significant decrease in acetol formation. Acetol was considered as the primary tar precursor during pyrolysis of biomass and lower concentration of acetol from torrefied biomass may generate lesser tars in the bio-oil. Torrefaction of biomass at 225°C followed by pyrolysis increased the amount of CO_2 formed relative to the control. However, higher torrefaction temperatures reduced the amount of CO_2 formed during pyrolysis. Concentration of water formed during torrefaction followed by pyrolysis was significantly less compared to control biomass.

Conclusions

Biomass torrefaction is the promising thermal pretreatment method for lignocellulosic biomass for thermo-chemical conversion platform. Torrefaction process significantly increases the energy density of biomass close to coal in addition to improve the storability and grindability of biomass. Higher torrefaction temperature had a significant effect on the increased energy density and decreased hemi-cellulose content (data not shown). Identification and separation of volatile compounds generated during torrefaction can become potential value added compounds in the future. Effect of torrefaction on the combustion, gasification and pyrolysis processes showed promising preliminary results on the final product compositions. Further research in this area at UGA is directed towards better understanding on the structural and functional modification of torrefied biomass and evaluating thermo-chemical reaction kinetics of both gasification and pyrolysis processes.

Acknowledgement

The author acknowledges the Traditional Industrial Program (TIP) of Georgia for Forest Resources and Products for funding this project and the invaluable research support from Dr. Jim Kastner & Mr. Gerry Lindo at the University of Georgia, Athens (UGA).

References

1. Boerrigter, H., S. V. B. Van Paasen, P. C. A. Bergman, J. W. Konemann, R. Emmen, A. Wijnands. (2005). "OLGA tar removal technology". *ECN Report # C-05-009*, The Netherlands: Energy Research Center of the Netherlands (ECN).
2. Dayton, D. (2002). "A review of the literature on catalytic biomass tar reduction". *NREL/TP report # 510-32815.* National Renewable Energy Laboratory, Golden, CO.
3. Mani, S., L. G. Tabil and S. Sokhansanj. (2004). "Grinding performance and physical properties of wheat and barley straws, corn stover and switchgrass". *Biomass and Bioenergy* 27(4): 339-352
4. Mani, S., L. G. Tabil and S. Sokhansanj. (2006a). "Effects of compressive force, particle size and moisture content on mechanical properties of biomass pellets", *Biomass and Bioenergy,* 30(7): 648-654.
5. Mani, S., S. Sokhansanj, X. Bi and A. Turhollow. (2006b). "Economics of producing fuel pellets from biomass", *Applied Engineering in Agriculture*, 22(3): 1-6
6. Prins, M. J., K. J. Ptasinski, F. J. J. G. Janssen. (2006b). "More efficient biomass gasification via torrefaction". *Energy* 31: 3458-3470.
7. Prins, M.J.; K.J. Ptasinski.; F.J.J.G. Janseen. (2006a). "Torrefaction of wood: Part 1. Weight loss kinetics". *J. Anal. Appl. Pyrolysis.* 77: 28-34.
8. Sokhansanj, S., A. Turhollow, J. Cushman and J. Cundiff. (2002). "Engineering aspects of collecting corn stover for bioenergy". *Biomass and Bioenergy* 23: 347-355.
9. Tillman, D. A. (2000). "Biomass co-firing: the technology, the experience, the combustion consequences". *Biomass and Bioenergy* 19(6): 365-384.

Simulation of Combustion and Gasification of Biomass Including Agricultural and Forest Residues and Pulp Mill Black Liquor Using FactSage

Hua-Jiang Huang, Department of Bioproducts and Biosystems Engineering, University of Minnesota, Saint Paul, MN

Shri Ramaswamy, Department of Bioproducts and Biosystems Engineering, University of Minnesota, Saint Paul, MN

There is an increasing interest to convert biomass such as agricultural waste, forest residue, and black liquor solids from pulp mills into heat, power and biofuels and biochemicals.

In conventional Kraft pulp mills, the black liquor solids containing organic chemicals such as hemicellulose and lignin, and inorganic chemicals (Na2S and Na2CO3 used for pulping) is currently burnt in the Tomlinson recovery boiler for recovery of the inorganic chemicals and generation of energy via combustion. The combustion with CHP (combined heat and power) system is not very energy-efficient. Instead, IGCC (Integrated Gasification Combined Cycle) with gasification replacing combustion can bring higher efficiency and environmental benefits. In addition, part of syngas produced from gasification can be used to produce biofuels and chemicals by syngas synthesis (e.g., Fisher -Tropsch synthesis) and conversion.

The combustion/gasification of black liquor is a very complicated process as it involves multiphase reactions including inorganic solid phases. In this study, the thermochemical software Factsage is used to simulate the combustion/gasification of biomass including black liquor solids and other agricultural or forest biomass, in order to better understand the combustion/gasification process and predict its performance or efficiency. Potential chemical emissions were also predicted to evaluate the environmental impact. The results can be useful for techno-economic and environmental analysis of the whole biomass-based biorefinery.

See more of this Session: Thermochemical Biorefineries II
See more of this Group/Topical: Forest and Plant Bioproducts Division

Catalytic Steam Reforming of Water Extracted Bio-Oil Fractions

Satya K. Jujjuri, Center for Sustainable Environmental Technologies, Iowa State University, Ames, IA

Justinus A. Satrio, Center for Sustainable Environmental Technologies, Iowa State University, Ames, IA

Robert C. Brown, Center for Sustainable Environmental Technologies, Iowa State University, Ames, IA

Steam reforming of bio-oil over a Ni-catalyst mainly generates syngas which can further be converted into a range of fuels and chemicals. Moreover, under suitable conditions and through catalyst modifications, product selectivity toward highly desirable hydrogen can be significantly improved by promoting water-gas shift reaction. Production of syngas from bio-oil is highly endothermic requiring higher temperatures while water-gas shift reaction that produces hydrogen is exothermic preferring lower temperatures. Thus optimum reaction conditions are required for improving yield toward syngas/hydrogen. Although steam reforming of petroleum-derived feedstocks is well established but because of the higher reactivity and thermal instability of bio-oil several challenges are faced in the fixed bed reformer wherein catalyst deactivation is rapid with bio-oils. Typically this condition is minimized by the utilization of higher steam to carbon ratios. The focus of this work is to study the catalytic conversion efficiency of different fractions of bio-oil that were obtained during the fast pyrolysis of biomass mainly in terms of the yield of syngas/hydrogen from wheat straw bio-oil. One of the main goals of this work is to identify the most desirable bio-oil fraction for the reforming step. A commercial Ni/Al2O3 catalyst is employed for testing. High steam to carbon ratios of 10 or higher are maintained by selectively extracting water soluble compounds from the four different bio-oil fractions produced during fast pyrolysis of wheat straw. The temperature range studied is between 350°C and 550°C at atmospheric pressure. Effluent gases are analyzed on a Micro-GC.

See more of this Session: Thermochemical Biorefineries II
See more of this Group/Topical: Forest and Plant Bioproducts Division

Review and Analysis of Transportation Fuel Production From Biomass-Derived Syngas

Christopher M. Kinchin, National Renewable Energy Laboratory, Golden, CO

Thermochemical conversion of biomass to synthesis gas (syngas), followed by catalytic conversion of syngas to hydrocarbons is one potential avenue for producing renewable liquid transportation fuels. The Biomass Program at the National Renewable Energy Laboratory (NREL) has performed techno-economic analyses for producing clean syngas and converting the syngas to mixed alcohols. However, the syngas can be catalytically converted to other hydrocarbons with higher energy density and potentially better infrastructure compatibility.

The Massachusetts Institute of Technology School of Chemical Engineering Practice recently worked with NREL to compile and compare a list of liquid transportation fuels that can be produced from syngas. The fuels are compared and ranked according to criteria such as energy density, infrastructure compatibility, technology maturity, and environmental impact. An initial screening process reduced an extensive list to the four most promising options, which were selected for further evaluation. A literature review and techno-economic analysis was thus performed for each of these four options.

The resulting scoping analysis includes mass and energy balances for each process as well as economic evaluation using assumptions and cost data consistent with previous NREL techno-economic analyses on biomass to mixed alcohols processes. Yields and costs for each process are estimated and compared. Additional results of the project are identification of critical knowledge gaps and recommendations for future research and development.

See more of this Session: Thermochemical Biorefineries II
See more of this Group/Topical: Forest and Plant Bioproducts Division

Chemistry Modeling of Biomass Pyrolysis: Application to the CFD Simulation of a Laminar Entrained Flow Reactor

Perrine Pepiot, National Renewable Energy Laboratory, Golden, CO

Mark W. Jarvis, National Renewable Energy Laboratory, Golden, CO

Mark R. Nimlos, National Renewable Energy Laboratory, Golden, CO

A current limitation to the efficiency of biomass gasification is the high level of tars usually found in the product gas, causing fouling and hazardous waste. Costly cleaning treatments to control the tar content of the exiting gas are usually applied downstream. Another appealing approach would be to directly optimize the gasifier design and operating conditions to the biomass properties for minimal tar production. However, such an approach requires a better understanding of the detailed chemical and physical processes occurring simultaneously in the reactor, including mixing, heat and mass transfer, and chemical reactions. Computational Fluid Dynamics (CFD) offers a unique opportunity to incorporate fundamental results, for example from detailed chemical modeling, into a larger scale framework, allowing to gain some valuable insight on tar formation pathways.

In this work, an extensive experimental investigation of biomass pyrolysis in a laminar entrained flow reactor is used as a basis for the development and validation of a reactive CFD model. The three-dimensional, cylindrical configuration remains simple enough to allow for a relatively detailed description of the chemical processes occurring in the reactor. Particles considered in the experiments are finely milled to a maximum size of 180 µm, which is below the size of a typical cell structure. Therefore, intra-particle transport processes are neglected in the CFD model. Particles are assumed to remain spherical in shape as they shrink during conversion, allowing the use of well-characterized correlations for the drag and heat transfer models. Biomass density is kept constant, and a volumetric approach is adopted to evaluate the chemical source terms for the solid phase.

A preliminary chemical mechanism describing biomass volatilization and further conversion of the products into permanent gases and polycyclic aromatic hydrocarbons has been assembled from our previous work on soot formation and data available in the literature. To reduce the computational cost associated with solving the chemistry in CFD, automatic reduction techniques have been applied to this detailed mechanism containing several hundreds of species to obtain a low-order scheme able to correctly reproduce the dynamics of the full chemical system. Since tar formation is the focus of this study, emphasis in the reduced model has been placed on the accurate description of aromatic rings formation from small hydrocarbon species and their subsequent molecular growth. The obtained chemical scheme has been integrated into CFD, and simulation results have been compared to the experimental data collected in the LEFR for a wide range of reactor temperatures and residence times. An improved description of the early stages of biomass volatilization based on carefully chosen model compounds has been derived from the analysis of the computations and the principal component results obtained in the experimental work.

See more of this Session: Thermochemical Biorefineries II
See more of this Group/Topical: Forest and Plant Bioproducts Division

Modeling of a Continuous Two-Phase Bubble Column Reactor for the Production of Biodiesel From Free Fatty Acids

Richard A. Cairncross, Department of Chemical and Biological Engineering, Drexel University, Philadelphia, PA

Minhazuddin Mohammed, Chemical & Biological Engineering, Drexel University, Philadelphia, PA

Cory Melick, Chemical & Biological Engineering, Drexel University, Philadelphia, PA

A continuous reactor model has been developed to optimize the performance of a two-phase bubble column reactor for biodiesel production from oils with high free fatty acid content using acid catalyst. The model accounts for the esterification reaction kinetics, thermodynamics and mass transfer limitations associated with the dynamics of methanol and water between vapor and oil phases. The reactor model is a set of differential mass balances for each species in each phase and contains characteristic parameters which can be manipulated under steady state to investigate operating regime of the reactor. The model parameters are estimated based on transient model results from a set of parametric studies consisting of feed temperature; alcohol feed quality, feed flow rates, agitation intensity and baffle design. The model enables optimization of the reactor design and operating conditions with fewer experiments. A finite difference method has been employed to solve the model and predict both transient and steady-state species concentration profiles inside the reactor. The model will be extended to capture the effects of simultaneous esterification and transesterification reactions inside the same reactor so the model can be applied to a variety of feedstock. The model predicts conversion profiles observed in experiments with varying flowrates of oil and alcohol feeds. Also, the model predicts the effect of feed quality (e.g. presence of water in the feed alcohol) on overall reactor conversion.

See more of this Session: Developments in Biobased Alternative Fuels II
See more of this Group/Topical: Sustainable Engineering Forum

Sipho C. Ndlela, Iowa Energy Center-BECON, Iowa State University, Nevada, IA

Evaluation of the reaction conversion and energy consumption during biodiesel production when using a high shear mixer pilot scale reactor

Biodiesel is a renewable fuel that has the potential to lower vehicle sulfur emissions when used directly or in a blended form and can be synthesized from fat and oil sources. These sources include refined vegetable oil, crude vegetable oil, waste oil, and animal fat. The two main reaction routes in the production of biodiesel are the esterification of free fatty acids in oil and the transesterification of triglycerides. In each of these routes, several processing stages are employed depending on the purity of the oil. For instance, crude soy bean oil goes through pre-processing stages such as degumming, neutralization, and bleaching prior to transesterification, glycerine removal, and final purification. The pre-processing steps are omitted however, when high purity soy bean oil is used as feedstock. Although using refined oil is the preferred route, the increased demand for this type of feedstock has resulted in high prices which have encouraged biodiesel manufacturers and researchers to find alternative oil sources. In addition to the search for alternative sources, more research is being conducted to evaluate other biodiesel processing methods and technologies. Both issues are discussed in the current study. First, soy bean oil was processed by two different biodiesel production methods which compared a pilot scale high shear mixer reactor to a stirred batch reactor. The shear mixer was made up of three pairs of fine rotor and stator sets that were more effective in converting triglycerides into methyl esters. Results from this study suggested that a high shear mixer gave higher conversions for shorter reaction times.

Previous experiments comparing a batch reactor to a high shear mixer were conducted such that the amounts of methanol and catalyst added were at 6 moles of methanol and 0.5 wt% sodium methoxide catalyst respectively, based on the added oil. The next set of experiments will evaluate the reaction conversions by increasing and then lowering either the amounts of methanol or catalyst. In addition, the energy consumption of the high shear mixer will be evaluated and then compared to that of the batch system.

High purity soy bean oil was previously tested; now a pre-processed waste oil sample with a free fatty acid (FFA) percentage below 0.4 wt% will be tested and its results compared to that of soybean oil. After looking at the short reaction times of the high shear mixer, the next phase will be to evaluate reaction conversion of refined soy bean oil as compared to that of pre-processed waste oil.

Keywords: Transesterification, high shear mixer, biodiesel technologies, biodiesel purification, glycerine

Introduction

From 1999 to 2008, biodiesel production in the US has increased dramatically for various reasons including the use of incentives through government programs (e.g. tax credits), enhanced energy security, the potential to reduce the environmental impact of fossil fuels, and the fact that it can be synthesized from renewable sources including vegetable oils, animal fats, and recycled oils [1,2,3,4]. Stricter sulfur emission regulations further motivated the use of biodiesel as an additive that provides lubricant properties necessary for diesel engines [1,4]. Blending biodiesel with #2 diesel at levels ranging from 5% to 20% without major engine modifications is common, but higher levels may require engine modifications [1,4,5]. Despite the benefits of biodiesel mentioned early, the presence of free glycerine and the higher cloud point can create engine problems during extreme winter temperatures [5,6]. Research and development continues to address cloud point issues by using additives or alcohols other than methanol at the end of biodiesel production. Biodiesel is typically synthesized either through an esterification or a transesterification reaction depending on the oil source. Esterification is achieved by reacting the free fatty acids in oil with an alcohol in the presence of an acidic homogeneous or solid catalyst. The transesterification route involves reacting a triglyceride with methanol to form methyl esters and glycerine using a base as catalyst. Transesterification is the preferred route since it eliminates pre-processing steps but requires the use of refined oil initially instead of poor quality oil with a high FFA%. Due to high demand, the price of high purity oil is high, resulting in the closure of biodiesel production facilities [1]. In order to tackle the problem of high feedstock prices, manufacturers and researchers are continuing to evaluate other oil rich crops including algae, jathropa, sunflower, and also other oil sources such as waste oil, grease, and used oil [1, 6]. In addition to the search for alternative sources, more research is being conducted to evaluate other biodiesel processing methods and technologies to lower operating costs [2,7]. Examples of biodiesel processing technologies that have been studied include using continuous stirred tank reactors (CSTR), ultrasonic reactors, enzymatic synthesis, supercritical methanol high pressure reactors, and high shear mixers [2,7].

During biodiesel synthesis, the mass transfer effects have a negative impact on the biodiesel conversion rates. These mass transfer effects can be eliminated by increasing mechanical agitation and mixing rates [7, 8]. Most studies typically focus on evaluating mixing rates using either stirred batch reactors or CSTRs [3-6,8]. However, mass transfer affects other types of systems including ultrasonic reactors, in which it is dependent on the ultrasonic cavitation rates [2], and supercritical methanol reactors, where it can be eliminated above the critical temperature and pressure [9]. A study by Noureddini et al. [7] evaluated reaction conversion of vegetable oils into biodiesel using a high shear mixer, which can be smaller in sized than stirred reactors and is very effective in making biodiesel at shorter reaction times. The current study will be looking at comparing the conversion rates of a high shear mixer to a stirred batch reactor. In addition, direct energy conversion is being evaluated for each reactor by monitoring the electricity consumption.

After biodiesel synthesis using a high shear mixer, the final step involves cleaning the crude biodiesel to remove salts, free glycerine, and to neutralize the base catalyst. The most common method of cleaning crude biodiesel is to neutralize it with an acid and then wash it in water. Neutralizing the catalyst with acid can create an unsafe work environment, and requires expensive construction materials to handle the acidic streams in the process. Also, high water consumption during biodiesel cleaning is a concern and may create a waste stream that is costly to dispose of. Using solid resins is an alternative method that can eliminate the need for water and acid. Currently, a purification solid resin: Amberlite BD 10 dry (supplied by Rohm and Hass) is being tested. Numerous analyses were conducted to evaluate the effect of BD 10 dry during the crude biodiesel purification process. These included testing for soap content, free fatty acids, as well as free/total glycerine. The total glycerine analysis gave an indication of how well the initial refined oil was converted into biodiesel.

Experimental setup and methods

Materials:
Refined soy bean oil purchased from ADM is the main feedstock tested throughout this study. Methanol (reagent grade) and sodium methoxide (25% concentration) was purchased from Sigma Aldrich. Crude biodiesel purification performed with Amberlite BD 10 dry (received from Rohm and Hass). The methanol to oil molar ratio was 6:1, and the sodium methoxide was added at 0.5 wt% based on the initial oil weight.

High shear mixer:
Experiments conducted in the pilot scale process using an IKA Inc. high shear mixer consisting of a fine rotor and stator set. In this study, each set of rotor and stator will be referred to as a generator. Tests using the high shear mixer were conducted using single, double, and triple generators. The single generator was run at 60 Hz, but when the double and the triple generators were added, the system power failed, so the double and triple generators were run at 50 Hz. Oil was first added to the reactor, mixed, and circulated through the heat exchanger until the temperature reached 50 – 55°C. At this point, a mixture of methanol and sodium methoxide was added and mixed for 3 – 5 minutes and a sample was then taken. The high shear mixer was started and the process was run by circulating the reactor contents for 60 minutes while maintaining the temperature at about 63°C. Samples were taken at 15 minute intervals.

Laboratory batch reactor:
At the laboratory, refined soy bean oil at 50°C was mixed with a solution of methanol and sodium methoxide. A 3 liter glass reactor was heated using a heating mantle and mixed with an overhead stirrer. The experiment was conducted at 60 – 63°C, for 60 min and samples were taken at 15 minute intervals.

Crude biodiesel neutralization and purification:
Crude biodiesel was cleaned and neutralized using commercially available ion exchange resin Amberlite BD 10 dry. The purpose of this step is to purify crude biodiesel by removing free glycerine, neutralizing the catalyst, and trapping salts. About 10 wt% resin per gram of oil was mixed for 20 minutes with crude biodiesel in a 3 liter reactor at 55 – 60°C. At the end of this treatment, resins and biodiesel were separated prior to drying under vacuum at 125°C and 150 mmHg . Data obtained from the laboratory purification is being used to develop a fixed bed reactor system that will be connected to the high shear mixer and used to purify crude biodiesel directly after its final glycerine separation stage.

Figure 1. Current biodiesel high shear mixer system

Results and discussion

Laboratory batch reactor:
The transesterification reaction was first studied using a 3 liter batch reactor in the laboratory. The reaction was conducted by first preheating soybean oil to 50 – 55°C, followed by adding 80% of the methanol/sodium methoxide mixture and further heating to 60 – 63°C, where it was reacted for an additional hour. The reactants were transferred into a separatory funnel and allowed to settle for about one hour. Glycerine was drained out and the second transesterification reaction begun by adding the remaining 20% of the methanol/sodium methoxide mixture. At the end of the second stage, the crude biodiesel was further treated to remove free gylcerine and contaminants.
The crude biodiesel purification system was set up by placing the Amberlite BD 10 resin in the reactor and mixing for 20 minutes. A typical biodiesel process utilizes an acid to neutralize the sodium methoxide followed by water washing to clean crude biodiesel before vacuum drying [10,11,12].

Results from the laboratory batch process are shown in Figure 2 below. As seen in the diagram, the amount of the total glycerine present decreases over the duration of the experiments. These results suggested that triglycerides converted to methyl esters through diglycerides, monoglycerides, and finally methyl esters. A number of studies have confirmed the transesterification routes suggested by Figure 2 [2-6,10,11]. At the end of the first transesterification stage, the amount of un-reacted triglycerides ranged between 5 – 10%. The unreacted triglycerides and intermediates were further reacted during the second stage of transesterification. These results are consistent with observations from other studies where vegetable oils were converted into methyl esters [2-12].

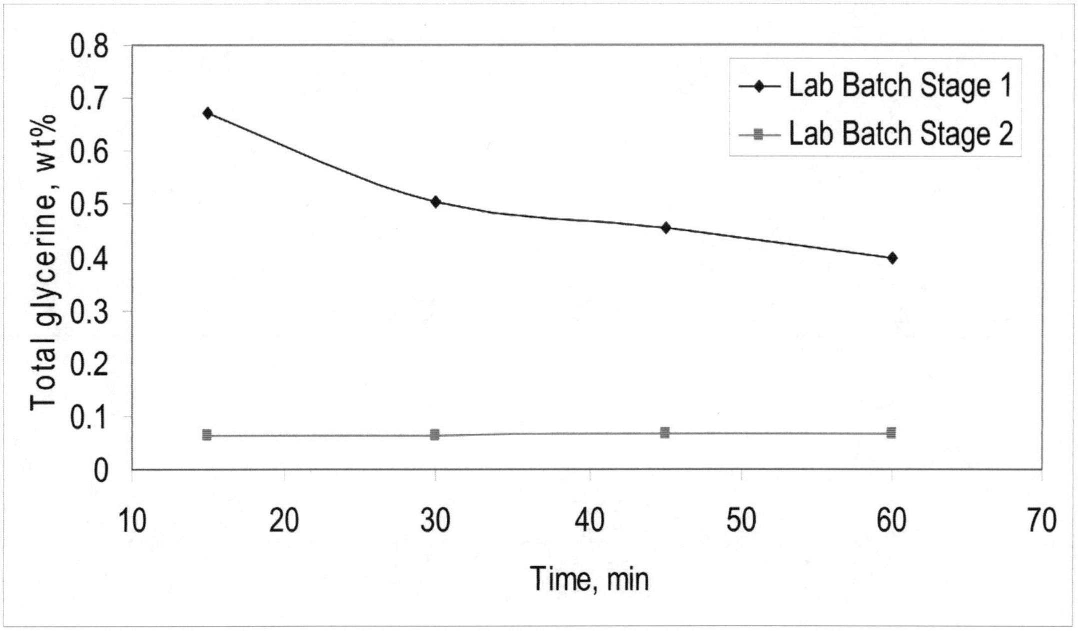

Figure 2. Transesterification reaction of refined soy bean oil into methyl esters using a laboratory batch reactor

High shear mixer experiments:
After evaluating the conversation rates using a batch type laboratory reactor, the next step was to scale-up the transesterification reaction using a high shear mixer which was designed for a capacity of 1 gallon per minute. The configuration of the mixer included a single mixing head generator (stator/rotor). The reaction was divided into two stages. The theoretical conversion for stage 1 was calculated to be above 90 wt%, whereas the remaining unreacted triglyceride was reacted in stage 2. Figure 3 shows results from both the high shear mixer and laboratory batch reactors. Based on these results the high shear mixer consisting of a single generator performed poorly when compared to the laboratory batch. The initial conversion rates were slightly higher for the laboratory batch reactor. Faster initial rates were observed when the double and triple generators were added. Repeated experiments using a single generator gave similar results. When comparing the batch reactor results to that of the single generator, similar conversion rates were noticed. In the case of double and triple generators, the conversion rates were higher than all other systems until the end of 45 minutes reaction time, after which the rates decreased. Overall, the double and triple generators gave consistent results throughout the experiment.

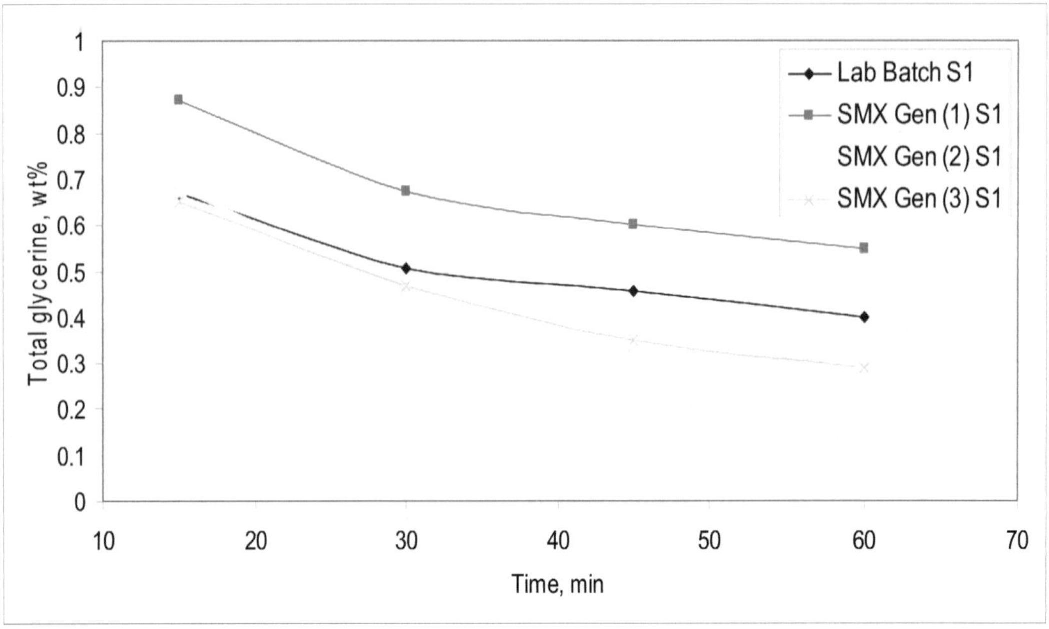

Figure 3. Transesterification reaction of refined soy bean oil into methyl esters using a laboratory batch reactor and high shear mixer, stage 1

Results from the second stage of transesterification are shown in Figure 4. Here, the single generator had a very low initial reaction rate from the start of the experiment till 15 minutes. The triple generator also had a low conversion rate. Both the batch reactor and the double generator gave consistent results with much higher conversion rates as compared to the other systems. The lowest values of total gylcerine ranged from 0.05 – 0.1 wt% as compared to 0.3 wt% for the single generator.

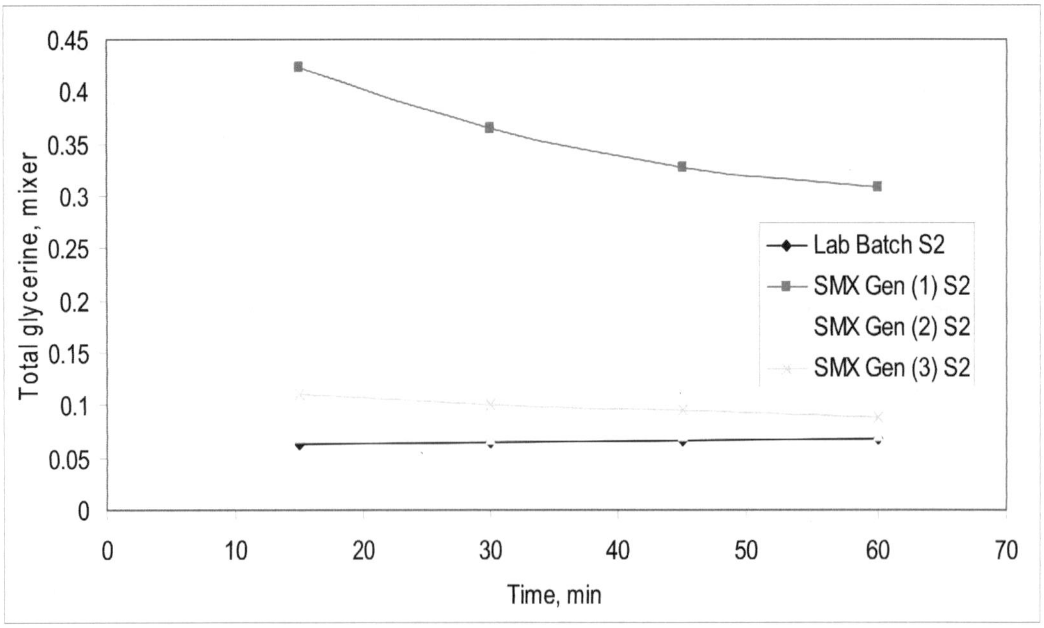

Figure 4. Stage 2 analysis results high shear mixer vs. a laboratory batch reactor.

Crude biodiesel purification:
After the second stage of transesterification experiments and glycerine separation, the crude biodiesel was purified using Amberlite BD 10 dry. Results of this process are shown in Figure 5. The data monitors the levels of free glycerine before and after exposure to Amberlite BD 10 dry for the triple generator for crude biodiesel after stages 1 and 2. As noted in Figure 5, the levels of free glycerine are higher before being subjected to the resins. After purifying with the resins, the amount of free gylcerine remaining is almost zero. From these results it is clear that the Amberlite BD 10 dry is effective in removing free glycerine. Other important factors to consider when using this resin is how well it neutralizes the catalyst and how it affects soap content. The pH of crude biodiesel was found to be 8 – 10, but after the purification stage using resins it ranged from 6 – 7. Soap and free fatty acid analysis showed varying results depending on the system being tested. In Table 1, results can be found showing the soap content of crude biodiesel prior to treatment and the acid number measured post-treatment.

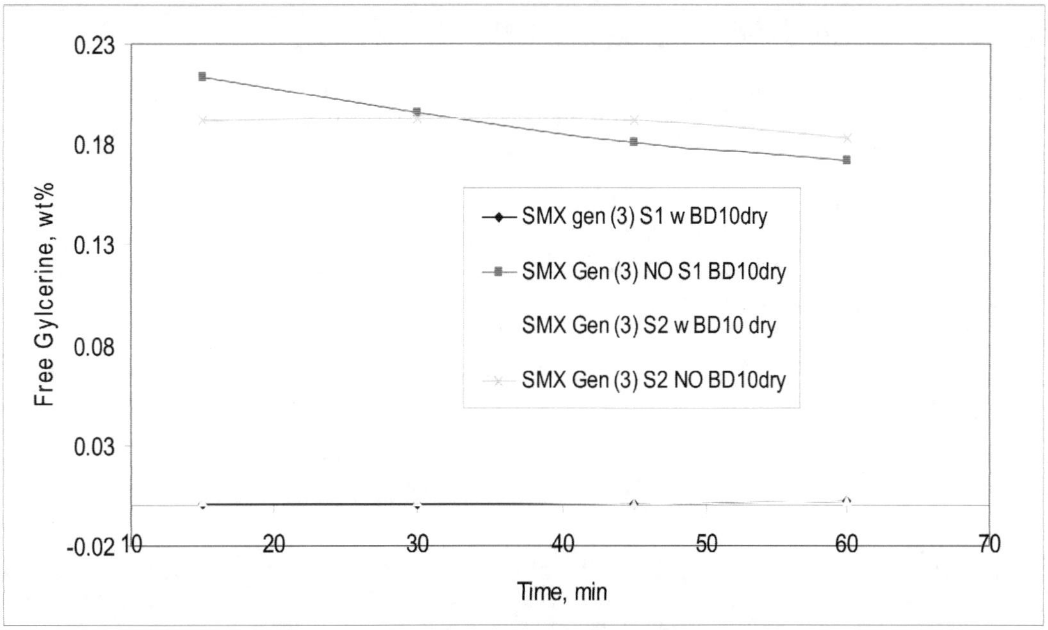

Figure 5: Effect of BD10dry in removing free gylcerine from crude biodiesel

As can be seen in Table 1, the soap levels vary and are mostly higher after the second transesterification stage. When comparing data for the double and triple generators in Table 1 to results in Figure 4, contradicting results are observed. For the double generator in Figure 4, the total glycerine was much lower than that of the triple generator but the soap level (from Table 1) for the double generator was almost twice that of the triple generator. In theory, increased soap levels may inhibit the formation of methyl esters, however, this behavior was not observed experimentally in this situation.

Using the Amberlite BD 10 dry resin to neutralize the catalyst and soap was important, but when excess soap was present, it was converted back into fatty acids. Such high levels of fatty acids expressed as acid number as seen in Table 1 ranged from 0.339 –

0.754 mgKOH/g depending on the sample. Note: acid numbers above 0.5 mgKOH/g exceed the B100 ASTM specification.

Name	Crude Biodiesel: soap content, ppm as Sodium Oleate	Final Biodiesel: Acid number, mgKOH/g
Lab Batch S1	263.423	0.339
Lab Batch S2	668.343	0.661
SMX Gen (1) S1	426.160	0.754
SMX Gen (1) S2	435.687	0.694
SMX Gen (2) S1	685.049	0.331
SMX Gen (2) S2	1115.129	0.349
SMX Gen (3) S1	671.900	0.349
SMX Gen (3) S2	783.604	0.358

Table 1: Effect of Amberlite BD 10 dry on soap content and acid number

Summary and work in progress

1. Shear mixer system with 2 stator/rotor set (double generator) suggested a better performance when converting soy bean oil into biodiesel

2. Current shear mixer study at 50 and 60 Hz; future work will look at lower ranges (10 – 40 Hz)

3. Optimize fixed beds for crude biodiesel purification

4. Integrate shear mixer study into a continuous biodiesel reactor system

5. Determine energy consumption of the shear mixer system

References

1. Proceedings from the Biodiesel Sustainability Symposium, November 19-20, 2008, St. Louis, Missouri, Organized by the, National Biodiesel Board Sustainability Task Force.

2. Naresh N. Mahamuni and Yusuf G. Adewuyi, *Optimization of the Synthesis of Biodiesel via Ultrasound-Enhanced Base-Catalyzed Transesterification of Soybean Oil using a Multifrequency Ultrasonic Reactor*, Energy and Fuels, March 5, 2009, A-J

3. Janulis, P., *Reduction of energy consumption in biodiesel fuel life cycle*, Renewable Energy 29, 2004, 861-871.

4. Gerhard Knothe, Analyzing Biodiesel: Standards and Other Methods, JAOCS, Vol. 83, no.10, 2006, 823.

5. Anton A. Kiss, Alexandre C. Dimian, Gadi Rothenberg, *Solid Acid Catalyst for Biodiesel Production-Towards Sustainable Energy,* Adv. Synth. Catal. 2006, 348, 75-81.

6. Knothe, G., J. Van Gerpen, and J. Krahl (eds.), *The Biodiesel Handbook*, AOCS Press, Champaign, Illinois, 2005.

7. Noureddini H., Harkey D., and Medikonduru V., *A Continuous Process for the Conversion of Vegetable Oils into Methyl Esters of Fatty Acids*, JAOCS, Vol. 75, no.12, 1998, 1775.

8. Noureddini H., Zhu D., *Kinetics of Transesterification of Soybean Oil*, JAOCS, Vol. 74, no. 11, 1997, 1457.

9. Demirbas Ayhan, *Biodiesel fuels from vegetable oils via catalytic and non-catalytic supercritical alcohol transesterifications and other methods: a survey,* Energy Conversion and Management 44, 2003, 2093-2109.

10. Barnwal B. K., Sharma M. P., *Prospects of biodiesel production from vegetable* oils in India, Renewable and Sustainable Energy Reviews 9, 2005, 363-378.

11. Zhang Y., Dube M. A., McLean D. D., Kates M., *Biodiesel production from waste cooking oil: 1. Process design and technological assessment*, Bioresource Technology 89, 2003, 1-16.

12. Zhang Y., Dube M. A., McLean D. D., Kates M., *Biodiesel production from waste cooking oil: 2. Economic assessment and sensitivity analysis*, Bioresource Technology 90, 2003, 229-240.

Acknowledgements

Project funded by the Iowa Energy Center under the BECON fund.

Software and process control by Van Meter Industrial, Des Moines, Iowa.

Optimization of Ultrasound-Enhanced Intensification of Biofuel Synthesis Using a Multifrequency Reactor and Taguchi Statistical Experimental Design Methodology

Naresh N Mahamuni and Yusuf G. Adewuyi*
Chemical Engineering Department
North Carolina Agricultural and Technical State University
Greensboro, NC 27405 USA

* To whom correspondence should be addressed. Phone: (336)334-7564, ext 107. Fax: (336)334-7417. E-mail: adewuyi@ncat.edu.

Introduction:

Biodiesel produced by base-catalyzed transesterification of vegetable oil is usually performed in batch reactors where the required energy is provided by heating accompanied by mechanical mixing. Using this technique, the best yields (up to 92-98%) are obtained using a methanol/oil molar ratio of 6:1, potassium hydroxide as the catalyst (1%), and at 65°C for one hour[1]. Alternatively, ultrasonication can provide an effective way to attain the required mixing while providing the necessary activation energy. It is well known that transesterification by ultrasound offer a lot of advantages over the conventional classical procedure[1]. It is found to be efficient (biodiesel yield up to 98-99%), as well as time and energy saving (dramatic reduction of reaction time to 5 min, compared to one hour or more using conventional batch reactor systems, and a remarkable reduction in static separation time to 25 min, compared to 8 h)[1]. Higher reaction rates, lower reaction time, higher yields, lesser reagents and better byproduct are some of the characteristics of the ultrasonic biodiesel synthesis processes[2,3]. Although ultrasound has been found to intensify the biodiesel formation, the science of biodiesel formation and relative importance of various parameters using ultrasound is not exactly known. Hence there is a need to understand the effect of various parameters such as frequency, power, oil/methanol mole ratio and catalyst on biodiesel synthesis in the presence of ultrasound. Traditionally, experiments have been done using the one variable at a time (OVAT) technique to study the effect of various parameters on the result of the process to achieve optimum conditions. In this technique, experiments are done by varying only one parameter at a time while all other parameters are kept constant. This results into a number of experiments which is time consuming, material consuming and is laborious. For example, to study the effect of four parameters (frequency, power, catalyst loading and molar ratio) at three different levels, 3^4 different combinations of parameters are possible. Also it is very difficult to identify and quantify the interactions among different parameters. Hence there was an absolute need for a design of experiments strategy which can reduce the number of experiments as well as identify and quantify the interactions among different parameters affecting the process. Dr. Genichii Taguchi[4] designed a system of specific orthogonal arrays to be chosen and applied in suitable conditions to describe a large number of experimental situations. The design of experiments using these orthogonal arrays by Dr. Taguchi is called 'Taguchi Design of Experiments Method'. In this method, the number of experiments is minimized while keeping all the information/data intact. The quantified and comparative analysis of the effect of parameters is the second advantage of this approach. Usually, with the aid of range analysis, analysis of variance (ANOVA) or analysis of signal-to-noise ratio (S/N ratio), the key factors that have significant effects on a response can be identified and the best factor levels for a given process can be determined from the pre-determined factor levels. Taguchi methodology uses several design arrays, such as L4, L8, L9, L12, L16, L18, L27 and L64, which focuses on the main effects and increases the efficiency and reproducibility of small-scale experiments. This is a quick yet accurate way of determining optimization. Details of the method can be found in the book by Roy[5].

Design of experiments:

Effect of four operating parameters/factors (ultrasonic power, ultrasonic frequency, catalyst (KOH) loading and soybean oil/methanol molar ratio) on biodiesel yield was considered at three levels as shown in Table 1. The factors and their levels were chosen according to a literature review on previous publications[2, 6,7] as well as our experience with the process[3,8].

Table 1. Selected parameters/factors and their levels

Level \ Factor	Frequency, kHz F	Power, W P	Catalyst loading (wt/wt) C	Molar ratio M
1	581	46	0.25	1:4
2	611	79	0.50	1:6
3	1300	143	0.75	1:9

For Taguchi-design of experiments with four factors and three levels for each factor, a standard L9 orthogonal array[9] was employed as shown in Table 2. Each row of the matrix represents one run at specified condition. To avoid the systematic bias, the sequence in which these runs were carried out was randomized[9]. No interactions among different parameters were assumed for this preliminary study. We are currently extending this study to include the effect of interactions among parameters.

Table 2. L9 Array for Design of Experiment by Taguchi Method

Exp. No.	Operating parameters			
	Frequency	Power	Catalyst loading	Molar ratio
1	1	1	1	1
2	1	2	2	2
3	1	3	3	3
4	2	1	2	3
5	2	2	3	1
6	2	3	1	2
7	3	1	3	2
8	3	2	1	3
9	3	3	2	1

Experiments were designed and carried out according to experimental conditions in Table 2.

Method of analysis:

The description of experimental set up, procedure and analysis is given in our earlier publication[3]. Biodiesel yield was measured using FTIR as described in our earlier publication[10].

Results and discussion:

The results of all measurements are summarized in Table 3. The biodiesel yields at 20 minutes were considered for this study in order to minimize the amount of energy input using ultrasound.

Table 3. The measured response (biodiesel yield) for each experiment

Exp. No	Frequency F	Power P	Catalyst loading C	Molar ratio M	Biodiesel Yield Y
1	581	46	0.25	1:4	7.89
2	581	79	0.50	1:6	71.69
3	581	143	0.75	1:9	94.39
4	611	46	0.50	1:9	5.58
5	611	79	0.75	1:4	82.78
6	611	143	0.25	1:6	60.59
7	1300	46	0.75	1:6	61.67
8	1300	79	0.25	1:9	8.34
9	1300	143	0.50	1:4	76.99

In Taguchi method, the results are statistically analyzed using analysis of variance (ANOVA) to determine the partial contribution of each operating factor on the response. The strategy of ANOVA calculation is to statistically analyze the variation that each factor causes relative to the total variation observed in the results[9,11]. This statistical table also screens the significant factors from those with less significance. The results of the ANOVA are shown in Table 4. As the percentage contribution of ultrasonic frequency was less than 2% (data not shown), factors with small variance were pooled to increase the statistical significance of important factors, as per recommendations of Roy[5]. It can be seen from Table 4 that the predominant factor that affects the biodiesel yield is the catalyst loading. Catalyst loading is followed by the ultrasonic power and molar ratio. Ultrasonic frequency has negligible effect on the biodiesel yield in the range of the study. Figure 1a shows that the average values of the biodiesel yield (response) at each level increased with increase in the ultrasonic power level. Similar trend is also observed in case of catalyst loading as shown in Figure 1b. Figure 1c and Figure 1d show the effects of molar ratio and ultrasonic frequency respectively. Using the "bigger is better" quality characteristics[5], the study of main effects, shown in Figure 1, indicate that the optimum condition is F1 P3 C3 M2, i.e. 581 kHz ,143 W , 0.75 % KOH (w/w) , 1:6 soybean oil/methanol mole ratio.

Table 4. ANOVA table for biodiesel yield (with pooling)

Factor	Degree of freedom f	Sum of squares S	Variance V	Variance ratio F	Pure sum of squares s'	Percent contribution P
Frequency	--	POOLED	--	--	--	--
Power	2	3967.511067	1983.755533	6.577517282	3364.318267	39.95435887
Catalyst	2	4227.114067	2113.557033	7.007898746	3623.921267	42.5686607
Molar ratio	2	1132.290267	566.1451333	1.877161443	529.0974667	11.40259747
ERROR	2	603.1928	301.5964	1	0	6.074382956
Total	8	9930.1082	--	--	--	100

The main effects of factors are determined using average values of response[5] at each level as shown in Figure 1.

Figure 1. Main Effects (without interactions)

(a)

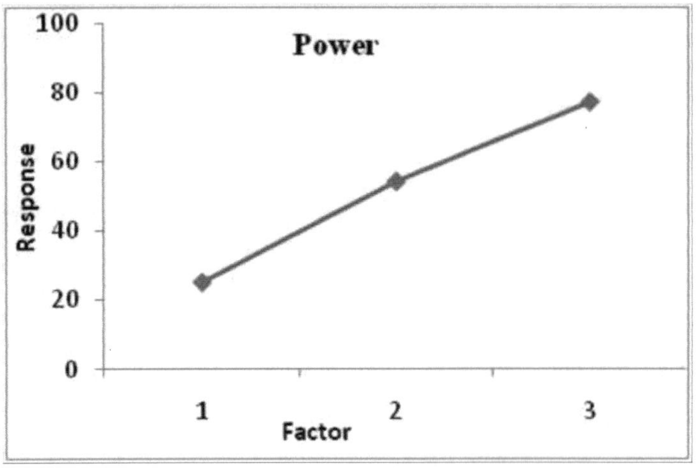

(b)

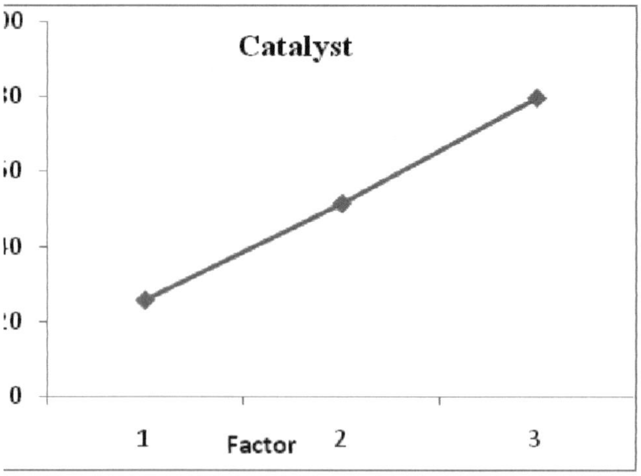

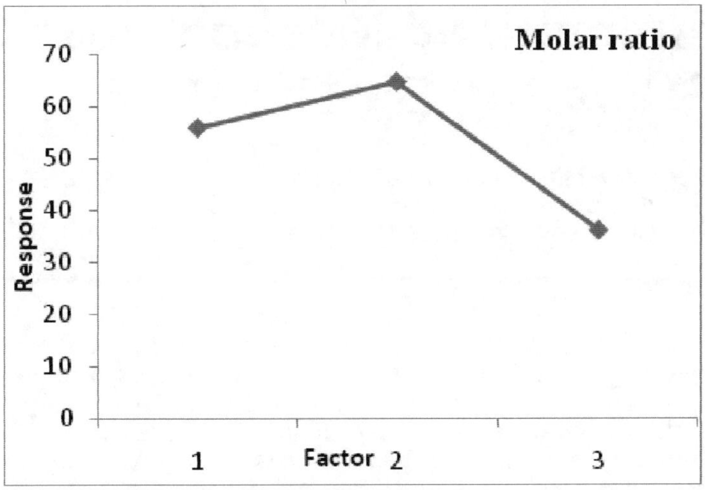

(c)

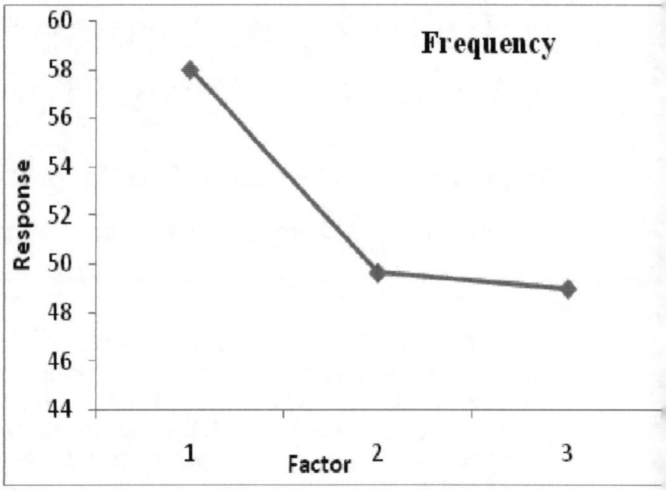

(d)

References:

1. Refaat, A. A., Elsheltawy, S. T. , Comparing three options for biodiesel production from waste vegetable oil *WIT Transactions on Ecology and the Environment* **2008,** *109* (Waste Management and the Environment IV), 133-140.

2. Stavarache, C.; Vinatoru, M.; Nishimura, R.; Maeda, Y., Fatty acids methyl esters from vegetable oil by means of ultrasonic energy. *Ultrasonics-Sonochemistry* **2005,** *12* (5), 367-372.

3. Mahamuni, N.; Adewuyi, Y., Optimization of the Synthesis of Biodiesel via Ultrasound-Enhanced Base-Catalyzed Transesterification of Soybean Oil Using a Multifrequency Ultrasonic Reactor. *Energy & Fuels* **2009,** *23* (5), 2757-2766.

4. Taguchi, G., *Introduction to quality engineering*. UNIPUB/Kraus International, White Plains: 1986.

5. Roy, R., *A primer on the Taguchi method*. Van Nostrand Reinhold, New York: 1990.

6. Ma, F.; Hanna, M., Biodiesel production: a review. *Bioresource Technology* **1999,** *70* (1), 1-16.

7. Meher, L.; Vidya Sagar, D.; Naik, S., Technical aspects of biodiesel production by transesterification—a review. *Renewable and Sustainable Energy Reviews* **2006,** *10* (3), 248-268.

8. Adewuyi, Y.; Oyenekan, B., Optimization of a Sonochemical Process Using a Novel Reactor and Taguchi Statistical Experimental Design Methodology. *Ind. Eng. Chem. Res* **2007,** *46* (2), 411-420.

9. Roy, R., *Design of experiments using the Taguchi approach: 16 steps to product and process improvement*. Wiley-Interscience: 2001.

10. Mahamuni, N.; Adewuyi, Y., Fourier Transform Infrared Spectroscopy (FTIR) Method To Monitor Soy Biodiesel and Soybean Oil in Transesterification Reactions, Petrodiesel Biodiesel Blends, and Blend Adulteration with Soy Oil. *Energy & Fuels* **2009,** *23* (7), 3773–3782.

11. Montgomery, D., *Design and analysis of experiments*. John Wiley & Sons New York: 2001.

Efficiency of Lipase Immobilized Mesoporous Silica as Catalyst for Preparing Biodiesel

Sang-Hyuk Ye, Chemical Engineering, Chung-Ang University, Seoul, South Korea

Won-Hyun Jeon, Chemical Engineering, Chung-Ang University, Seoul, South Korea

Sun-Geon Kim, Chemical Engineering, Chung-Ang University, Seoul, South Korea

Four different kinds of lipase (lipase PS/lipase AK)-immobilized catalysts ere prepared: meso (MCM-41) and meso/macroporous silica modified with and without octyltetraethoxysilane (OTES). They were used to prepare biodiesel from a commercial soybean oil at room temperature and 1 atm. The meso/macroporous silica with OTES resulted in higher yield, reducing the formation of tetracosanoic acid methyl ester significantly. The addition of water to transesterification reactor in small amount considerably extended the life of the catalysts. The yield, composition and life of the catalysts were compared with those of commercial lipase-immobilized catalysts such as CalB and lipase-immobilized Toyonite.

See more of this Session: Developments in Biobased Alternative Fuels II
See more of this Group/Topical: Sustainable Engineering Forum

Optimization of a Pilot Plant for Production of Biodiesel From Jatropha Curcas L. Oil

Rogelio Sotelo-Boyás, Instituto Politécnico Nacional (National Polytechnic Institute of Mexico), Mexico City, Mexico

Ma. del Pilar Piña-Delgado, Instituto Politécnico Nacional (National Polytechnic Institute of Mexico), Mexico City, Mexico

Morayma Uribe-Gomez, Instituto Politécnico Nacional (National Polytechnic Institute of Mexico), Mexico City, Mexico

An optimized process on pilot plant scale was carried out for the production of biodiesel from Mexican jatropha curcas L. oil. Initial studies were carried out at lab scale in order to determine the experimental conditions that could lead to a high yield of biodiesel from the alkaline transesterification of jatropha oil. The experimental conditions were in between the following intervals: temperature: 50 to 65 °C, reaction time: 30 to 120 min., catalyst type: NaOH and KOH, molar ratio of methanol to oil: 6:1 to 12:1, weight molar ratio of catalyst to oil: 0.5 to 1.5%. The optimal conditions were then used in the processing of the oil in a pilot plant. The plant consists of two 50 L-reactors that can be operated in either batch or continuous mode. Prior reaction, the oil was pretreated to reduce the amount of free fatty acids and moisture. After reaction, the biodiesel and glycerol phases were purified in two separated distillation towers. The optimization of the pilot plant was also carried out by simulating the plant with ASPEN PLUS. The yield of methyl esters at optimal conditions was 84wt%. The biodiesel was also characterized and its properties were found to be within the ASTM specified limits. The optimization of the reaction process at lab scale along with the simulation and optimization of the process at pilot scale leaded to an innovative separation and reaction schemes for the production of a high quality biodiesel from jatropha curcas L. oil.

Kewwords: jatropha, pilot plant, optimization, simulation.

See more of this Session: Developments in Biobased Alternative Fuels II
See more of this Group/Topical: Sustainable Engineering Forum

Conversion of Inexpensive Oils Into Biodiesel Using ZnO Nanoparticle Catalyst

Shuli Yan, Wayne State University, Detroit, MI

Manhoe Kim, National Biofuel Energy Laboratory, NextEnergy, USA, Detroit, MI

Steven Salley, Chemical Engineering and Materials Science, Wayne State University, Detroit, MI

Simon K. Y. Ng, Chemical Engineering and Materials Science, Wayne State University, Detroit, MI

Biodiesel is a mixture of fatty acid esters which can be produced from vegetable oils or animal fats with methanol. Recently, many researchers focus on using waste or unrefined oils as feedstock to decrease the production cost of biodiesel. A two-step method was reported, which firstly esterified free fatty acids (FFA) with methanol in the presence of H2SO4, then transesterified oil with methanol in the presence of NaOH. This production process is long and highly corrosive. In this study, a single-step method was developed using Zn-La mixtures as catalyst to convert crude soybean oil, crude coconut oil, crude palm oil, crude corn oil from DDGs, crude algae oil, waste cooking oil and the well-refined soybean oil with 3 % water and 5 % oleic acid additions into FAME. Oil transesterification and FFA esterification reactions simultaneously took place in one step. Effects of catalyst structure, metal oxide molar ratio, FFA and water contents in feedstock, reaction temperature and oil concentration on the yield of biodiesel were investigated. In a batch stir reactor, 42:1 molar ratio of methanol to oil, 2.3 (wt) % of catalyst, a high yield (95 %) of biodiesel was obtained within 3 hours using the above oils as feedstock. In a continuous reactor, 26:1 molar ratio of methanol to oil, 200 °C and 100 ~ 450 Psi, the yield of FAME maintained at 94 % for 49 days. This class of zinc and lanthanum mixture catalysts expands the oil feedstock for biodiesel production.

See more of this Session: Developments in Biobased Alternative Fuels II
See more of this Group/Topical: Sustainable Engineering Forum

Controlling the Rheology of Biomass

J. R. Samaniuk, Chemical and Biological Engineering, University of Wisconsin, Madison, WI

C. Tim Scott, USDA Forest Products Laboratory, Madison, WI

T. W. Root, Chemical and Biological Engineering, University of Wisconsin, Madison, WI

D. J. Klingenberg, Chemical and Biological Engineering, University of Wisconsin, Madison, WI

Rising global demand for energy combined with an increasing appreciation for the environmental consequences of non-renewable, petroleum based fuel use is lending urgency to the need for a carbon-neutral, widely accessible source of renewable energy. Particularly critical to the transportation industry is an abundant source of renewable liquid fuels such as ethanol to counter chronically rising petroleum costs. A likely feedstock in the production of fuel ethanol is lignocellulosic biomass. The greatest advantage of biomass as a feedstock is that it is the most abundant renewable carbon resource on the planet. The National Renewable Energy Laboratory (NREL) has already proposed a process for converting lignocellulosic biomass to liquid fuel. The process has tremendous promise but the economics of it are inhibitory to large scale implementation. Currently, NREL's process calls for using roughly 10 wt% biomass at any step in the conversion. The utility and equipment costs associated with heating and pumping of 90 wt% water slurries can be dramatically reduced by increasing the solids fraction. NREL's plans call for an increase in solids content from 10 to 40 wt% resulting in a 4-fold decrease in heating energy cost per pound of dry biomass and a 6-fold decrease in the water requirement.

Processing of high solids fraction biomass leads to a number of difficulties that are not present in more dilute systems. Concentrated biomass acts as a Bingham fluid with a yield stress that increases rapidly with solids concentration. This behavior makes pumping and transporting the process slurry difficult if not impossible. In addition, the mass transfer required in order to mix reactants, enzymes and reagents into the slurry is severely inhibited. Controlling the rheology of the biomass feed is critical for the success of high solids fraction biomass processing.

In this presentation, we describe the use of rheological modifiers to control the rheology of biomass slurries. We have found that by using high molecular weight water-soluble polymers, we can significantly reduce the yield stress of wood fiber and corn stover slurries. Using only a few wt% (based on the dry solid biomass) of these rheological modifiers, yield stresses can be reduced several-fold, permitting the flow of biomass at high concentrations. We will describe the effects of properties of the rheological modifiers (e.g., type, concentration and molecular weight) on rheological behavior, show evidence of the microscopic mechanisms producing the observed behavior, and examine the effects of the modifiers on the downstream biomass processing (e.g., acid hydrolysis, enzymatic hydrolysis and fermentation).

See more of this Session: Biomass Refining: Unit Operations, Processes, and Optimization
See more of this Group/Topical: Forest and Plant Bioproducts Division

Impact of Hemicellulose Pre-Extraction for Bioconversion On Birch Kraft Pulp Properties

David Hodge, Chemical Engineering & Materials Science, Michigan State University, East Lansing, MI

Jonas Helmerius, Chemical Engineering, Luleå University of Technology, Luleå, Sweden

Jonas Vinblad von Walter, Sun Pine Biodiesel AB, Piteå, Sweden

Curt Lindström, Smurfit Kappa Kraftliner AB, Piteå, Sweden

Ulrika Rova, Chemical Engineering, Luleå University of Technology, Luleå, Sweden

The carbohydrate portion of lignocellulosic feedstocks are ideally suited to conversion via biochemical transformations because of their crucial role in cellular metabolism. The combination of hemicelluloses extraction with pulping processes could be one way to generate a sugar feedstock amenable to biochemical transformation to fuels and chemical intermediates. White liquor, green liquor, and water HC extractions of birch wood were performed under conditions compatible with the Kraft process, at different times, temperatures and alkali charges. The effective alkali charge was in extractions between 0%-7% and temperature between 130°C-160°C for 20-90 minutes. The xylan yields from different HC extractions were measured and the chips from select HC extractions were cooked, and the refined pulps were made into hand sheets. Several metrics for hand sheet quality were compared with a reference pulp made from the same wood chips. It is possible using white liquor to extract xylan from birch wood chips prior Kraft cooking without decreasing the pulp yield and paper strength properties, and at the same time achieve an impregnation of the wood chips. It is not possible in that extraction to attain extracted and hydrolyzed liquor containing a fermentable concentration of xylose, 2.63 g/L in this study. Increased extracted wood material, increased final acetic acid concentration and decreased final xylan concentration together with increased effective alkali charge at the same extraction temperature and time in white liquor extractions performed support that xylan degradation increases. Using white liquor or green liquor under the conditions investigated degrades xylan resulting in significant losses of xylose that could have been used as substrate in fermentation processes.

See more of this Session: Biomass Refining: Unit Operations, Processes, and Optimization
See more of this Group/Topical: Forest and Plant Bioproducts Division

Kinetics of Sulfur Deactivation of Steam Reforming Catalyst

Stefan Czernik, National Bioenergy Center, National Renewable Energy Laboratory, Golden, CO

Whitney Jablonski, National Bioenergy Center, National Renewable Energy Laboratory, Golden, CO

We are developing a gasification process to produce clean syngas from biomass feedstocks. The syngas can be further processed to produce fuels and chemicals such as gasoline, diesel, alcohols, olefins, synthetic natural gas, and high purity hydrogen. The raw biomass-derived syngas contains unwanted components such as methane, other gaseous hydrocarbon and polycyclic aromatic compouds (collectively called tars), which must be removed prior to further processing. Steam reforming using nickel-based catalysts is the preferred method to remove these impurities and convert them to additional amounts of carbon oxides and hydrogen. The main challenge in the tar removal process is the presence of hydrogen sulfide, a known catalyst poison, in biomass-derived syngas. Depending on the biomass feedstock, the syngas produced in our pilot plant contains 20-600 ppm H2S and steam reforming catalysts loose activity after a relatively short time on stream (30-50 minutes). In support of our catalyst development effort, we conducted a catalyst deactivation and regeneration study using a fixed-bed micro reactor system (0.5-2.0 g catalyst) with model tar compounds (ethylene and benzene). This work reports on the results of the catalyst activity during steam reforming of ethylene containing different amounts of H2S. The product gas composition was measured at the reactor outlet every three minutes during the test, which allowed us to determine conversion of ethylene as a function of time on stream. When the H2S concentration in ethylene was 100 ppm or higher, the catalyst performance measured as ethylene conversion decreased very rapidly and the sampling frequency was too low to provide reliable time-conversion data for kinetics calculation. At lower H2S concentration the loss of the catalyst performance was slow enough and allowed an attempt to determine the deactivation kinetics calculation. The experimental data are well described by a catalyst deactivation model with residual activity, which will be presented in this paper.

See more of this Session: Biomass Refining: Unit Operations, Processes, and Optimization
See more of this Group/Topical: Forest and Plant Bioproducts Division

Increasing the Economic Efficiency of a Biomass Feedstock Used for Greenhouse Heating

Monty Singletary, Dave C. Swalm School of Chemical Engineering, Mississippi State University, Mississippi State, MS

Mark Bricka, Dave C. Swalm School of Chemical Engineering, Mississippi State University, Mississippi State, MS

Commercial greenhouses, while not considered a traditional staple of the forest products industry, can provide a stable outlet for waste biomass as a heat source. Originally using natural gas, the Natchez Trace Greenhouse in Kosciusco, MS, has recently converted to using alternative fuels for its heating needs. These greenhouses use biomass as a source of supplemental heat during the winter months, and the amount of land present at the site presents ample possibility for large quantities of biomass storage during the warmer months. Initially, corn was used as a fuel source, however, the cost of corn in Mississippi has risen almost parallel to that of natural gas. Waste sawdust from a local mill is available as a very cost effective fuel, but it alone does not provide adequate heating necessary for the number of greenhouses present on the property. This study aims to find an amendment that can be added to the waste wood to allow for ample heating, while meeting the greenhouse's monetary constraints.

See more of this Session: Biomass Refining: Unit Operations, Processes, and Optimization
See more of this Group/Topical: Forest and Plant Bioproducts Division

The earth's finite fossil fuel reserves and the increase in demand for energy have raised many concerns that have led researchers to focus on the development of new technologies for the production of sustainable and renewable energy. Lignocellulosic bioethanol is one of the promising alternative sources of energy, as it is produced from renewable materials that do not directly compete with food and can be grown on non-agricultural lands. The process of producing bioethanol requires three main steps: pretreatment, enzymatic hydrolysis, and fermentation. Pretreatment is considered to be the most important step because it dictates the susceptibility of the biomass to the downstream processes, hydrolysis and fermentation.

The purpose of pretreatment is to alter the structure of the biomass so that cellulose, which is entrapped in the lignin and hemicellulose matrix, can become more amenable to enzymatic deconstruction. Some of the desired characteristics of a pretreatment are: high recovery of carbohydrates, modification of lignin, hydrolysis of hemicelluloses, minimum sugar degradation, and allowing high digestibility of cellulose by cellulases. Ethanol organosolv pretreatment has been shown to be effective on softwoods and hardwoods because it addresses these requirements.

In this pretreatment, the material is treated in an aqueous-organic solvent mixture with the addition of an inorganic acid catalyst, H_2SO_4, and heated at high temperatures for 40 – 60 min. After pretreatment, three fractions are recovered: Ethanol Organosolv Lignin (EOL) fraction, solid fraction, and water-soluble fraction. The lignin that is dissolved in the organophilic phase is referred to as EOL; the solid fraction contains mainly cellulose and some residual lignin and hemicelluloses; and the water-soluble fraction has the soluble lignin, monomeric and oligomeric hemicelluloses.

The focus of this research is to better understand the organosolv pretreatment chemistry by exploring the fundamental characteristics of pretreated biomass in the context of developing an efficient bioconversion of cellulose to glucose. Pretreatment of *Buddleja davidii* was performed at three conditions (A, B, and C), which produced substrates with different compositions and/or enzymatic hydrolysis profiles. The most sever condition, condition C caused more delignification and carbohydrate hydrolysis/degradation than conditions A and B. Results showed higher %EOL, higher % furfurals, and lower Klason lignin (KL) content in the solid fraction for condition C, when compared to the other two conditions. Glucose recovery was ~85% of the original glucose in the wood for the three conditions.

However, although the wood in condition C was delignified more than in condition B, ~19% KL and ~9% KL for substrate B and C respectively, their enzymatic hydrolysis profiles were similar. Both of these substrates were readily digestible by cellulases. This was interesting because it is believed that lignin has a large impact on the efficiency of enzymatic hydrolysis. A set of experiments were conducted in order to explain this behavior. The idea was to study the structure of cellulose, crystallinity of cellulose, and its degree of polymerization (DP), which are another three important factors that affect enzymatic hydrolysis. Gel-permeation chromatography (GPC) was used to determine the DP; the ultrastructure and crystallinity index (CrI) of cellulose were investigated using solid-state CP/MAS ^{13}C NMR, in terms of the relative amounts of cellulose I_{α}, cellulose I_{β}, *para*-crystalline cellulose, and celluloses at accessible and inaccessible surfaces.

Results showed that efficient organosolv pretreatment was capable of reducing DP and CrI of cellulose, and convert the stable crystalline cellulose dimorphs (I_α/I_β) to the more reactive *para*-crystalline and amorphous celluloses. These outcomes made the cellulose more accessible to enzymatic attacks due to two reasons. First, lower DP of cellulose means higher number of reducing chain ends of cellulose, which is an enhancement for efficient enzymatic deconstruction, because cellobiohydrolase enzyme acts mainly on the reducing chain ends of cellulose. It is one of the main enzymes used to convert cellulose to glucose besides endoglucanase and cellobiase. Second, shorter chains of cellulose and less crystalline cellulose cause the formation of weaker networks that lack strong intermolecular hydrogen bonding, allowing for easier digestibility by cellulases.

In this study, a plant called *Buddleja davidii* was used as a resource for bioethanol production because it has unique properties to become an agro-energy crop (*1*). *B. davidii* exhibits a very wide range of growth habitat and is well adapted to growing in poor soil conditions (*1*). The plant is perennial, has moderate growth dimensions, and is susceptible to very few pests or diseases (*1*).

We have assessed this new bioresource by determining its composition and by elucidating the chemical structures of both lignin and cellulose. Results from compositional analysis indicate that *B. davidii* has low ash and extractives content, low cellulose (35%), high hemicellulose (34%), and relatively high lignin content (30%) (*1*). As mentioned above, the ultrastructure and crystallinity index (CrI) of cellulose were investigated using solid-state CP/MAS ^{13}C NMR, in terms of the relative amounts of cellulose I_α, cellulose I_β, *para*-crystalline cellulose, and celluloses at accessible and inaccessible surfaces. The predominant crystalline form of cellulose is *para*-crystalline cellulose (33%) and the crystallinity index (CrI) is determined to be 0.55 (*1*). The CP/MAS ^{13}C NMR spectrum and cellulose signal assignments are presented in **Figure 1** (*1*). The region in the NMR spectrum of cellulose between δ 80 and 92 is the most informative region, which is assigned to C-4 carbons. The area between δ 86 and 92 represents the crystalline and *para*-crystalline forms of cellulose, whereas the broader signal between δ 80 and 86 corresponds to the amorphous domains (*1*). The various ultrastructures of cellulose were determined using an analysis method based on a non-linear least-square fitting of CP/MAS ^{13}C NMR spectra (*1*). **Figure 2** shows the spectral fitting for the C-4 region of the CP/MAS ^{13}C NMR spectrum of cellulose (*1*).

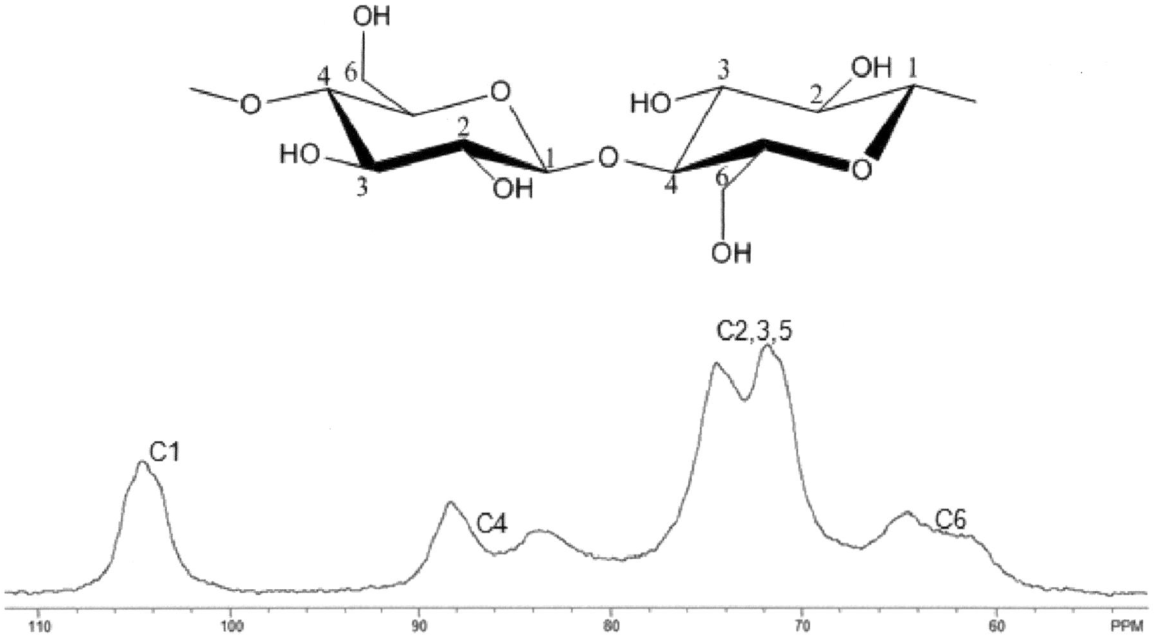

Figure 1. Solid-state CP/MAS ^{13}C NMR spectrum of cellulose (*1*).

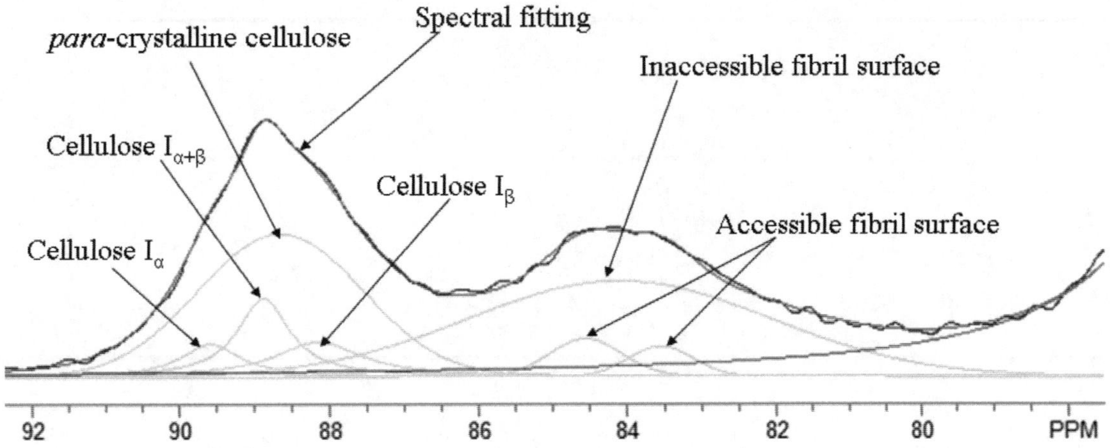

Figure 2. Spectral fitting for the C-4 region of CP/MAS ^{13}C NMR spectrum of cellulose (*1*).

Structural characterization of lignin has been performed using quantitative ^{13}C NMR and ^{31}P NMR spectroscopy. The quantitative ^{13}C NMR spectrum is presented in **Figure 3** (*1*). This technique provides a facile method to determine the amount of various functional groups and lignin structures. The spectrum regions for *p*-hydroxyphenyl (***h***) units, aromatic quaternary C, and aromatic tertiary C have δ-range of 162-157, 160-123, and 123-103 respectively (*1*). ^{13}C NMR reveals that the lignin in *B. davidii* is of the hardwood type with a syringyl to guaiacyl ratio of 19:81 (*1*). Quantitative ^{31}P NMR spectroscopy is a technique based on the derivatization of phenolic structures with 2–chloro–4,4,5,5–tetramethyl–1,3,2–dioxaphospholane (TMDP) and the use of internal standard (Cyclohexanol) in order to determine the amount of different hydroxyl functional groups (*1*). The quantitative ^{31}P NMR spectrum for *B. davidii* lignin is presented in **Figure 4** (*1*)**.** The dominant OH group in the lignin is at aliphatic sites (*1*). The lignin has considerable amount of free guaiacyl OH, but low carboxylic acid OH (*1*).

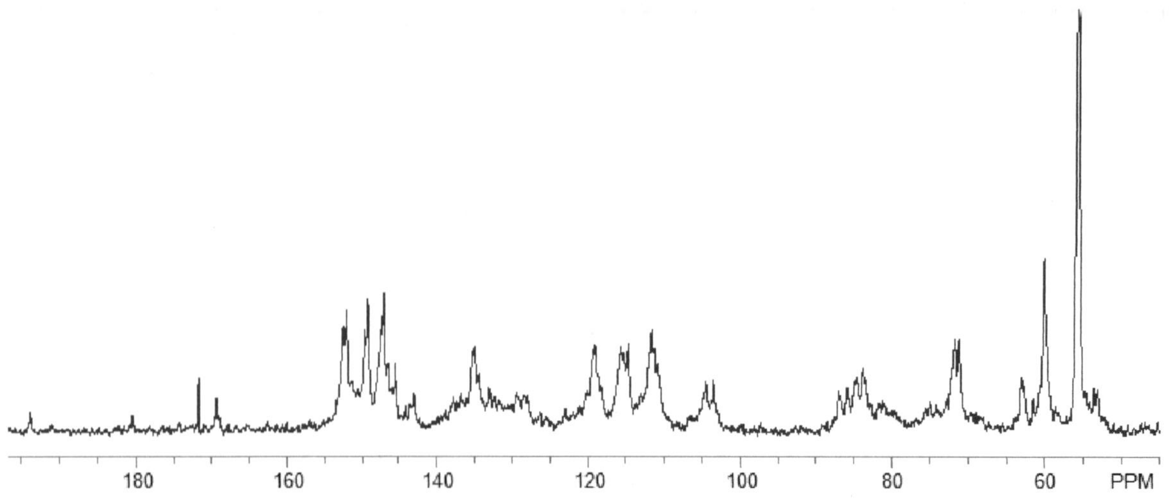

Figure 3. Quantitative ^{13}C NMR spectrum of lignin from *B. davidii* (*1*).

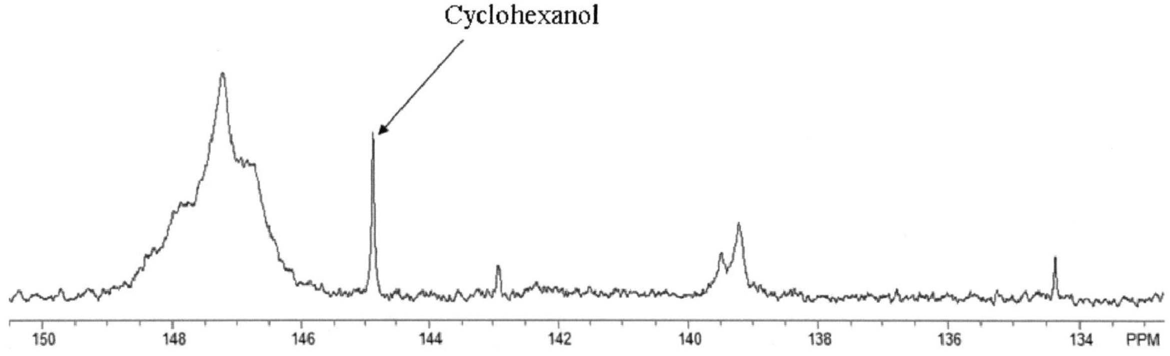

Figure 4. Quantitative ^{31}P NMR of lignin from *B. davidii*.

The characteristics of *B. davidii* can be categorized into attractive and undesired features. The positive ones include: the wide range of growth habitat, the favorable growth dimensions, and the low degree of polymerization of cellulose (1000 DP). The negatives ones include: the low cellulose content, the high hemicellulose and lignin contents, and the high crystallinity index. The high and low descriptions were based on comparing *B. davidii* to other softwoods and hardwoods. For instance, spruce has 27%

lignin, 41% cellulose, 31% hemicellulose, 0.45 crystallinity index, and 3300 DP. Aspen has 21% lignin, 48% cellulose, 27% hemicellulose, 0.47 crystallinity index, and 2500 DP.

In summary, the results of this study provide fundamental investigation of the biomass chemistry, the organosolv pretreatment technology, and its ability to prepare solid cellulosic biomass for conversion by enzymatic hydrolysis. This presentation also summarizes the attractive features and challenges associated with using *B. davidii* for bioethanol production.

1. Hallac, B. B.; Sannigrahi, P.; Pu, Y.; Ray, M.; Murphy, R. J.; Ragauskas, A. J., Biomass Characterization of *Buddleja davidii*: A Potential Feedstock for Biofuel Production. *Journal of Agricultural and Food Chemistry* **2009,** 57, 1275-1281.

Investigation and Modeling Natural Biodegradation System in Soil; Application for Designing An Efficient Biological Pretreatment Technology for Biofuel Production

Mythreyi Chandoor, Biological Systems Engineering, Washington State University, Pullman, WA

Deepak Singh, Biological Systems Engineering, Washington State University, Pullman, WA

Shulin Chen, Biological Systems Engineering, Washington State University, Pullman, WA

During the process of lignocellulosic biomass conversion to Biofuel, the greatest challenge now remains in the efficiency of conversion of the biomass into hydrolysable sugars. As the C-5 andC-6 sugar components are attached and protected by lignin matrix, getting the sugars depend primarily on the level of deconstruction of lignin. The natural biodegradation process of lignocellulosic biomass in the soil provides an excellent platform to investigate how the biomass is converted. The process is essential for maintaining the carbon-cycle in the natural atmosphere, which reflects that the study on the biomass degradation in soil can be a safest model to imprint the lignocellulosic degradation process for obtaining Biofuel.

It is a complex mechanism which involves soil as a support media for the growth and interaction of different sets of microcosm including algae, fungi and bacteria. This work presents the microcosm involved in the biodegradation processes occurring in the soil and detailed structural characterization of biomass reflecting the mechanism of biodegradation process. Understanding of the degradation mechanism in soil will provide important information to improve the current biomass pretreatment technology. In addition to this, the potential lignin degrading microbes/and or microcosm may be useful for the lignocellulosic biodegradation process as well as for the bioremediation of hazardous organic compounds.

In this study, we used wood chips and wheat straw as substrates in the soil. It has characteristics such as carbon content of 2.285 gm/Kg, sulfur content of 0.026 gm /kg and nitrogen content of 0.171 gm/kg. The 3 gms of lignocellulosic samples were incubated with 440 gms of soil at 20 oC for four months and the moisture level was maintained to 15% which issame as in natural condition. The sampling of the biomass was done in every four weeks. The presentation will include three different categories of study. The analysis of structural changes/modification of lignin and the sequential cleavage bond in the structure of lignin observed during the incubation of wood chips and wheat straw in soil for every set of four weeks until about twenty weeks by NMR, FTIR and GC-MS analysis. The second will be the isolation and characterization of the microcosm involved in the lignin degradation process in soil. In this, the strains were isolated from the biomass and characterized on the basis of their ability to decolorize Azo dye. Dye discoloration assay was observed in A647 nm after the strains were grown in LB media with dye concentration of 0.002% incubating at 28 oC for 24 hrs. Interestingly, some of the strains showed high discoloration activity within 16 hrs. The mechanism behind the discoloration and the strains identification is under investigation.

The modeling of the lignin degrading soil system is based on the physical and biochemical conditions in the soil would ultimately provide to design a novel method of pretreatment. As the process is taking place mainly due to the interaction between different sets of microcosm, our presentation would provide a new perspective of pretreatment technology.

See more of this Session: Biomass Refining: Unit Operations, Processes, and Optimization
See more of this Group/Topical: Forest and Plant Bioproducts Division

AUTHOR INDEX

AUTHOR INDEX

AUTHOR INDEX